Genomic Selection in Animals

Genomic Selection in Animals

JOEL IRA WELLER
Institute of Animal Sciences
Agricultural Research Organization
Bet Dagan, Israel

WILEY Blackwell

Published by John Wiley & Sons, Inc., Hoboken, New Jersey
Published simultaneously in Canada

For general information on our other products and services or for technical support, please contact our Customer Care Department within the United States at (800) 762-2974, outside the United States at (317) 572-3993 or fax (317) 572-4002.

Wiley also publishes its books in a variety of electronic formats. Some content that appears in print may not be available in electronic formats. For more information about Wiley products, visit our web site at www.wiley.com.

Library of Congress Cataloging-in-Publication Data

Names: Weller, Joel Ira, author.
Title: Genomic selection in animals / Joel Ira Weller.
Description: Hoboken, New Jersey : John Wiley & Sons Inc., [2015] | Includes bibliographical references and indexes.
Identifiers: LCCN 2015042390 | ISBN 9780470960073 (cloth)
Subjects: | MESH: Animals, Domestic–genetics. | Breeding. | Genetic Markers–genetics.
| Quantitative Trait Loci–genetics. | Selection, Genetic.
Classification: LCC SF105.3 | NLM SF 105 | DDC 636.08/21–dc23
LC record available at http://lccn.loc.gov/2015042390

Cover credit: Domestic animals © Jevtic/iStock/Getty Image Plus; Colorful smooth twist light lines background © VikaSuh/iStockphoto; Genes © Ingram Publishing/Getty Images

Set in 10.5/12pt Times by SPi Global, Pondicherry, India

2 2016

For Elisha Eliyahu, my special grandson

Contents

Preface: Welcome to the "Promised Land"

> ...And I saw a man who was standing in the gate. He looked as if he were bronze. In his hands, he had a string and a measuring stick.... The stick that the man had was 6 long cubits. But each cubit was a cubit plus the width of a hand. The man measured the wall. It was one stick high and one stick wide....
>
> Ezekiel 40: 3–5

I have been involved in the field of genetic markers and quantitative trait loci since I began my doctorate under the direction of Prof. Morris Soller and Dr. Thomas Brody in 1977. In my doctorate thesis we grew 2000 tomato plants and used morphological and biochemical markers (isozymes). Since the early 1980s, Dr. Soller was convinced that marker-assisted selection was "just around the corner." Now I can say without any exaggeration that we have arrived in the "promised land." Marker-assisted selection, now generally termed genomic selection, has become a reality over the last 5 years for most of the important farm animals, especially dairy cattle. However, genomic evaluation is still very much a "work in progress." Although there is definitely sufficient material in the literature to justify a text of this nature for graduate students, I am quite sure that a similar text in 5 years will look quite different.

When writing a book of this nature, one is always confronted with the problem of what to assume is already known by the reader and what has to be explained. Generally with respect to biology, very little is required of the reader. Anyone with a B.A. or B.Sc. in biology should have no problem with any biological concepts presented. Specifically with respect to genetics, I am assuming that the reader has a basic understanding of quantitative genetics, such as could be obtained from the classic *Quantitative Genetics* of Falconer (1964), or *Genetics and Analysis of Quantitative Traits* by Lynch and Walsh (1998). With respect to mathematics, I am assuming that the reader is familiar with both differential and integral calculus and has a basic familiarity with matrix algebra. Applications of matrix algebra specific to animal breeding are explained in some detail, even though this information has become quite standard for any graduate student in applied genetics. Detailed explanation of the physics and chemistry of current technologies used to genotype large numbers of markers and whole genome sequencing is outside the scope of this book.

Finally I want to thank those people who made this book possible. I have already mentioned my teachers Morris Soller and Thomas Brody, and I would also add the late Ram Moav and Reuven Bar-Anan. Also I thank my colleagues both in Israel and the United States, especially Micha Ron, Ephraim Ezra, Ignacy Misztal, George Wiggans, Paul VanRaden, and John Cole. I also thank my editors Justin Jeffryes and Stephanie Dollan, who I have yet to meet face-to-face, and last but not least my family, and especially my lovely wife Hedva, who has given me every support in this and in all my other endeavors.

Menachem Av, 5775

1 Historical Overview

Introduction

Genomic selection is based on the synthesis of statistical and molecular genetics that occurred during the last three decades. In this introductory chapter we will review the landmark breakthroughs that lead to this synthesis. The first section reviews the milestones in the synthesis of Mendelian and quantitative genetics. The next section reviews the early experiments of quantitative trait locus (QTL) detection using morphological and biochemical markers, beginning with Sax's landmark experiment with beans (*Phaseolus vulgaris*). The following sections describe the development of DNA-level markers starting with restriction fragment length polymorphisms (RFLPs) to single nucleotide polymorphisms (SNPs) and copy number variations (CNV). The final sections describe QTL detection and marker-assisted selection (MAS) prior to genomic selection.

The Mendelian Theory of Genetics

Modern genetics is usually considered to have started with the rediscovery of Mendel's paper in 1900. The rediscovery of Mendel's laws led to a rapid first synthesis of genetics, statistics, and cytology. Boveri (1902) and Sutton (1903), first proposed the "chromosomal theory of inheritance" that the Mendelian factors were located on the chromosomes. Using *Drosophila*, Morgan (1910) demonstrated that Mendelian genes were linked and could be mapped into linear linkage groups of a number equal to the haploid number of chromosomes. Hardy (1908) and Weinberg (1908) independently derived their famous equation to describe the distribution of genotypes in a segregating population at equilibrium. That is, the frequencies of genotypes for a locus with two alleles with frequencies p and $1-p$ will be p^2, $2p(1-p)$, and $(1-p)^2$ for homozygotes for p-allele and heterozygotes and homozygotes for the other allele, respectively.

In 1919 Haldane derived a formula to convert recombination frequencies into additive "map units" denoted "Morgans" or "centimorgans," assuming a random distribution of events of recombination along the chromosome. The Haldane mapping function (Haldane, 1919) is based on the assumption of zero "interference" throughout the genome. That is, all events of recombination are statistically independent. In this case the number of events of recombination in any given chromosomal segment corresponds to a Poisson distribution. The map distance between

Genomic Selection in Animals, First Edition. Joel Ira Weller.

two genes in Morgans, M, which is a function of the frequency of observed recombination between them, R, is derived as follows:

$$M = -\frac{1}{2}\ln(1-2R) \tag{1.1}$$

The Mendelian Basis of Quantitative Variation

Unlike the morphological traits analyzed first by Mendel and then by Morgan, most traits of economic interest in agricultural species display continuous variation, rather than the discrete distribution associated with Mendelian genes. Despite the early synthesis between Mendelian genetics and cytogenetics, there seemed to be no apparent connection between Mendelian genetics on the one hand and quantitative variation and natural selection on the other.

Experiments by Johanssen (1903) with beans demonstrated that environmental factors are a major source of variation in quantitative traits, leading to the conclusion that the phenotype for these traits is not a reliable indicator for the genotype. Yule in 1906 first suggested that continuous variation could be explained by the cumulative action of many Mendelian genes, each with a small effect on the trait. (Many different terminologies have been employed for these genes. I will use the term "QTL" throughout.) Fisher in 1918 demonstrated that segregation of QTL in an outcrossing population would generate correlations between relatives. Payne (1918) demonstrated that the X chromosome from selected lines of *Drosophila* contains multiple factors, which influenced scutellar bristle number. Thus, by 1920, the basic theory necessary for detection of individual genes affecting quantitative traits was in place.

Detection of QTL with Morphological and Biochemical Markers

In 1923 Sax demonstrated with beans that the effect of an individual locus on a quantitative trait could be isolated through a series of crosses, resulting in randomization of the genetic background with respect to all genes not linked to the genetic markers under observation. Even though all of his markers were morphological seed markers with complete dominance, he was able to show a significant effect on seed weight associated with some of his markers.

During the next 50 years, there were relatively few successful experiments that found marker–QTL linkage in plant and animal populations, and of these even fewer were independently repeated. A major problem was the relatively small size of most experiments. In most cases in which QTL effects were not found, power was too low to find segregating QTL of a reasonable magnitude (Soller *et al.*, 1976).

In 1961 Neimann-Søressen and Robertson proposed a half-sib design for QTL detection in commercial dairy cattle populations. Although the actual results were disappointing, this was the first attempt to detect QTL in an existing segregating population. All previous studies were based on experimental populations produced specifically for QTL detection. This study was also groundbreaking in other aspects. It was the first study to use blood groups rather than morphological markers, and the proposed statistical analyses—a χ^2 (chi-squared) test, based on a squared sum of normal distributions, and ANOVA—were also unique. This was the first study that attempted to estimate the power to detect QTL and to consider the problem of multiple comparisons when several traits and markers were analyzed jointly.

Lewontin and Hubby showed in 1966 that electrophoresis could be used to disclose large quantities of naturally occurring enzyme polymorphisms in *Drosophila*. Almost all enzymes analyzed showed some polymorphism that could be detected by the speed of migration in an electric field. Studies with domestic plant and animal species found that electrophoretic polymorphisms were much less common in agricultural populations. During the 1980s there were a number of QTL detection studies in agricultural plants based on isozymes using crosses between different strains or even species in order to generate sufficient electrophoretic polymorphism (Tanksley *et al.*, 1982; Kahler and Wherhahn, 1986; Edwards *et al.*, 1987; Weller *et al.*, 1988). It was clear, though, that naturally occurring biochemical polymorphisms were insufficient for complete genome analyses in populations of interest.

DNA-Level Markers, 1974–1994

The first detected DNA-level polymorphisms were RFLPs. Grodzicker *et al.* (1974) first showed that restriction fragment band patterns could be used to detect genetic differences in viruses. Solomon and Bodmer (1979) and Botstein *et al.* (1980) proposed RFLP as a general source of polymorphism that could be used for genetic mapping. Although RFLPs are diallelic, initial theoretical studies demonstrated that they might be present throughout the genome. Beckmann and Soller (1983) proposed using RFLP for detection and mapping of QTL. The first genome-wide scan for QTL using RFLP was performed on tomatoes by Paterson *et al.* (1988). In animal species, however, RFLP markers were homozygous in most individuals and therefore have not been as useful for QTL mapping.

A major breakthrough came at the end of the decade with the discovery of DNA microsatellites. Mullis *et al.* (1986) proposed the "polymerase chain reaction" (PCR) to specifically amplify any particular short DNA sequence. Using the PCR, large enough quantities of DNA could be generated so that standard analytical methods could be applied to detect polymorphisms consisting of only a single nucleotide. Since the 1960s, it has been known that the DNA of higher organisms contains extensive repetitive sequences. In 1989 three laboratories independently found that short sequences of repetitive DNA were highly polymorphic with respect to the number of repeats of the repeat unit (Litt and Luty, 1989; Tautz, 1989; Weber and May, 1989). The most common of these repeat sequences was poly(TG), which was found to be very prevalent in all higher species. These sequences were denoted "simple sequence repeats" (SSR) or "DNA microsatellites."

Microsatellites were prevalent throughout all genomes of interest. Nearly all poly(TG) sites were polymorphic, even within commercial animal populations. These markers, unlike most morphological markers, were by definition "codominant." That is, the heterozygote genotype could be distinguished from either homozygote. Furthermore, microsatellites were nearly always polyallelic. That is, more than two alleles were present in the population. Thus, most individuals were heterozygous. Relatively dense genetic maps based on microsatellites were generated nearly in all agricultural species (e.g., Ihara *et al.*, 2004), and these markers were also used to detect and map segregating QTL. The weaknesses of microsatellites are twofold: First their distribution throughout the genome is not sufficiently dense for determination of causative polymorphisms responsible for observed QTL. (The causative polymorphisms will be denoted "quantitative trait nucleotides" (QTN).) Second, due to the repeat structure of microsatellites, PCR amplification was generally not exact, and "stutter bands" with varying numbers of the repeat unit were generated. Various rules were developed to estimate the actual genotype from the observed PCR product, but the analysis could not be fully automated. A technician still had to review each individual genotype, and error rates in genotype determination were in the range of 1–5%.

DNA-Level Markers Since 1995: SNPs and CNV

Since 2000 "SNPs" (reviewed by Brookes (1999)) have supplanted microsatellites as the marker of choice for genetic analysis. An SNP is generally defined as a DNA base pair location at which the frequency of the most common base pair is lower than 99%. Unlike microsatellites, which usually have multiple alleles, SNPs are generally biallelic, but are much more prevalent throughout the genome, with an estimated frequency of one SNP per 300–500 base pairs. In human populations differences in the base pair sequence of any two randomly chosen individuals occur at a frequency of approximately one per 1000 kb (Brookes, 1999). Thus, SNPs can be found in genomic regions that are microsatellite poor. SNPs are apparently more stable than microsatellites, with lower frequencies of mutation. Beginning in 2005, methods were developed for automated scoring of first thousands and then hundreds of thousands of microsatellites per individual. Genotyping error rates are in the range of 0.05–0.01% with "BeadChip" technology (Weller *et al.*, 2010). A detailed description of the technologies developed for high-throughput SNP analysis is beyond the scope of the current text. For details, see Matukumalli *et al.* (2009).

QTL Detection Prior to Genomic Selection

Generally both natural and commercial populations are at linkage equilibrium for the vast majority of the genome. The exception is genomic sites that are closely linked on the same chromosome. Unlike genetic linkage within families that extends over tens of centimeters, population-wide linkage disequilibrium (LD) extends in animals over less than 1 cM (Sargolzaei *et al.*, 2008; Qanbari *et al.*, 2010). Therefore, unless a segregating genetic marker is closely linked to a QTL segregating in the population with an effect on some trait of interest, no effect will be associated with the marker genotypes. Thus naturally occurring LD could not be exploited prior to the advent of high-density genome scans. To detect the effect of a single QTL in outbred populations prior to high-density genome scans, it was necessary to generate LD.

In an analysis of inbred lines we are confronted with the opposite problem. That is, a significant effect associated with a genetic marker may be due to many genes throughout the genome and not necessarily to genes linked to the genetic markers. In crosses between inbred lines it was necessary to devise an experimental design that isolates the effects of the chromosomal segments linked to the segregating genetic markers.

Experimental designs can be divided into designs that are appropriate for crosses between inbred lines and those designs that can be used for segregating populations. Most early analyses performed to detect QTL have been based on planned crosses, although studies on humans, large farm animals, and trees have used existing populations. For humans, most species of domestic animals, and fruit trees, it is impractical to produce the inbred lines. Instead, experimental designs were based on the analysis of families within existing populations. Three basic types of analyses have been proposed—the "sib-pair" analysis for analysis of many small full-sib families, the "full-sib" design for analysis of large full-sib families, and the "half-sib" or "daughter design" analysis for large half-sib families.

MAS Prior to Genomic Selection

Prior to genomic selection, two MAS breeding programs were initiated in dairy cattle based on microsatellites in German and French Holsteins (Bennewitz *et al.*, 2004b; Boichard *et al.*, 2006). Both programs computed marker-assisted genetic evaluations (MA-BLUP) based on the algorithm of Fernando and Grossman (1989).

In the German program, markers on three chromosomes were used. The evaluations were distributed to Holstein breeders who used these evaluations for selection of bull dams and preselection of sires for progeny testing. The algorithm only included equations for bulls and bull dams, and the dependent variable was the bull's daughter yield deviation (VanRaden and Wiggans, 1991; derivation and use of daughter yield deviations will be discussed in detail in Chapter 6). Linkage equilibrium throughout the population was assumed. To close the gap between the grandsire families analyzed in the German granddaughter design and the bulls in use in 2004, 3600 bulls were genotyped in 2002. Until 2008, about 800 bulls were evaluated each year. Only bulls and bull dams were genotyped, since tissue samples were already collected for paternity testing. Thus additional costs due to MAS were low, and even a very modest genetic gain could be economically justified. This scheme was similar to the "top-down" scheme of Mackinnon and Georges (1998) in that evaluation of the sons was used to determine which grandsires were heterozygous for the QTL and their linkage phase. This information was then used to select grandsons based on which haplotype was passed from their sires. It differed from the scheme of Mackinnon and Georges (1998) in that the grandsons were preselected for progeny test based on MA-BLUP evaluations (Fernando and Grossman, 1989), which include general pedigree information in addition to genotypes.

The French MAS program included elements of both the "top-down" and "bottom-up" MAS designs. Similar to the German program, genetic evaluations including marker information were computed by a variant of MA-BLUP, and only genotyped animals and nongenotyped connecting ancestors were included in the algorithm. Genotyped females were characterized by their average performance based on precorrected records (with the appropriate weight), whereas males were characterized by twice the "yield deviations" of their nongenotyped daughters (yield deviations will also be explained in Chapter 6). Twelve chromosomal segments, ranging in length from 5 to 30 cM, were analyzed. Regions with putative QTL affecting milk production or composition were assumed to be located on bovine chromosomes 3, 6, 7, 14, 19, 20, and 26; segments affecting mastitis resistance on chromosomes 10, 15, and 21; and chromosomal segments affecting fertility on chromosomes 1, 7, and 21. Each region was found to affect one to four traits, and on an average three regions with segregating QTL were found for each trait. Each region was monitored by two to four evenly spaced microsatellites, and each animal included in the MAS program was genotyped for at least 43 markers. Sires and dams of candidates for selection, all male AI ancestors, up to 60 AI uncles of candidates, and sampling daughters of bull sires and their dams are genotyped. The number of genotyped animals was 8000 in 2001 and 50,000 in 2006.

Guillaume *et al.* (2008) estimated by simulation the efficiency of the French program. Breeding values and new records were simulated based on the existing population structure and knowledge of the variances and allelic frequencies of the QTL under MAS. Reliabilities of genetic values of animals less than 1 year old obtained with and without marker information were compared. Mean gains of reliability ranged from 0.015 to 0.094 and from 0.038 to 0.114 in 2004 and 2006, respectively. The larger number of animals genotyped and the use of a new set of genetic markers can explain the improvement of MAS reliability from 2004 to 2006. This improvement was also observed by the analysis of information content for young candidates. The gain of MAS reliability with respect to classical selection was larger for sons of sires with genotyped daughters with records.

Summary

By 2005 dense genetic maps based on DNA-level genetic markers were developed for nearly all economically important animal species. Numerous studies demonstrated that QTL affecting traits of economic importance could be detected via linkage to genetic markers. Theory was developed for MAS based on selection of a relatively small number of chromosomal segments, and several MAS breeding programs for dairy cattle were implemented in two countries. The "rules of the game" were to change dramatically in 2006 with the development of high-throughput SNP chips, which will be discussed in detail in the next chapter.

2 Types of Current Genetic Markers and Genotyping Methodologies

Introduction

Although a detailed description of DNA technology is outside the scope of this book, a brief discussion of the types of markers that were used for marker-assisted selection and the markers currently used for genomic selection has been included, as the characteristics of these markers affect the methodologies that have been developed for marker-assisted and genomic selection. In the final section we briefly review the current state of complete genome sequencing, which in all likelihood is the "wave of the future."

From Biochemical Markers to DNA-Level Markers

As noted in the previous chapter, the first study to use biochemical markers (as opposed to morphological markers) to detect segregating QTL was the study of Neimann-Sørensen and Robertson (1961), which used blood groups as genetic markers. During the 1960s it became clear that there was considerable variation in enzyme sequence that could be detected by electrophoresis. A number of studies were concluded during the 1980s using electrophoretic markers to detect segregating QTL in plant species (e.g., Weller *et al.,* 1988). However, electrophoretic markers were not polymorphic in commercial animal species. In addition to blood group markers, polymorphisms were also found in milk proteins, and several studies were performed to detect QTL via linkage to these markers (e.g., Bovenhuis and Weller, 1994).

The first DNA-level genetic markers found in animal species were restriction fragment length polymorphisms (RFLP). Although several studies were performed in plants to detect QTL via linkage to RFLP (Paterson *et al.,* 1988), these markers were not found to be very polymorphic in domestic animal species. A major breakthrough occurred with the development of the polymerase chain reaction (PCR) (Mullis *et al.*, 1986). Via the PCR it was possible to specifically amplify any particular short DNA sequence, provided unique primer sequences could be constructed. Thus large enough quantities of DNA could be generated so that standard analytical methods could be applied to detect polymorphisms consisting of only a single nucleotide.

Genomic Selection in Animals, First Edition. Joel Ira Weller.

DNA Microsatellites

Since the 1960s it has been known that the DNA of higher organisms contains extensive repetitive sequences. In 1989 three laboratories independently found that short sequences of repetitive DNA were highly polymorphic with respect to the number of repeats of the repeat unit (Litt and Luty, 1989; Tautz, 1989; Weber and May, 1989). The most common of these repeat sequences was poly(TG), which was found to be very prevalent in all higher species. These sequences were denoted "simple sequence repeats" (SSR) or "DNA microsatellites." Microsatellites were prevalent throughout all genomes of interest. Nearly all poly(TG) sites were polymorphic in the number of TG repeats, even within commercial animal populations. These markers were by definition "codominant." That is, the heterozygote genotype could be distinguished from either homozygote. Furthermore, microsatellites were nearly always polyallelic. That is, more than two different alleles were present in the population. Thus, most individuals were heterozygous.

During the 1990s genotyping costs per polymorphism were reduced from approximately $10 per genotype to about 1$ per genotype, due to development of machines specifically designed for this purpose, specifically the ABI DNA sequencer, which implemented nonradioactive analysis methods. In addition costs were reduced due to multiplexing of PCR, which runs several PCR in the same sample, and improved software for analysis. Dense genetic maps based on microsatellites were generated for most agricultural species, and genome scans for segregating QTL were performed for most agricultural animal populations of interest (reviewed by Weller (2007)).

Despite these advantages, microsatellites had several significant drawbacks, due chiefly to the prevalence of "stutter bands." (These bands are generated by "mistakes" in DNA replication during the PCR, in which a unit of the repeat motif is either deleted or added. Thus instead of a single clear band for each allele, secondary bands are also generated.) First, although software was developed to determine genotypes from the banding pattern, genotyping could not be completely automated. It was still necessary for a qualified technician to review the software results and make corrections. Second, genotyping error rates were often unacceptably high (e.g., Weller *et al.*, 2004). Finally the average density of microsatellites in the genome was not sufficient to capture population-wide linkage disequilibrium, which we will see is now the basis of genomic selection.

Single Nucleotide Polymorphisms

Since 1995 new classes of markers have also come into use. Chief among them are "single nucleotide polymorphisms" (SNP) (reviewed by Brookes (1999)). An SNP is generally defined as a base pair location at which the frequency of the most common base pair is lower than 99%. Unlike microsatellites, which usually have multiple alleles, SNPs are generally diallelic, but are much more prevalent throughout the genome, with an estimated frequency of one SNP per 300–500 base pairs. In human populations differences in the base pair sequence of any two randomly chosen individuals occur at a frequency of approximately one per 1000 kb (Brookes, 1999). Thus, SNPs can be found in genomic regions that are microsatellite poor. SNPs are apparently more stable than microsatellites, with lower frequencies of mutation. Ranade *et al.* (2001) first described conditions for genotyping large numbers of individuals for any SNP and computational methods that allow genotypes to be assigned automatically.

Several companies developed genotyping platforms for high-throughput genotyping of tens and even hundreds of thousands of SNPs simultaneously. By 2008 genotyping costs for SNPs were reduced to below $0.01 per genotype and are currently approximately $0.002 per genotype.

Currently the leading technology for high-throughput SNP genotyping is "Infinium HD assay" (http://support.illumina.com/content/dam/illumina-support/documents/myillumina/67f59f89-51ee-44d6-b1bb-a53dcb5bd01e/infinium_hd_ultra_user_guide_11328087_revb.pdf). Based on this technology, "mid-density BeadChips" have been developed for all the major agricultural animal species including 50–60 thousand markers. The "BovineHD BeadChip" (Illumina, Inc., San Diego, CA) was developed which includes 777,000 SNPs that span the entire cattle genome. A poultry array with over 580,000 markers has also been developed and is commercially available (http://www.affymetrix.com/catalog/prod670010/AFFY/Axiom%26%23174%3B+Genome%26%2345%3BWide+Chicken+Genotyping+Array#1_1). "High-density" marker arrays with more than half a million markers are under development for other major agricultural species.

Copy Number Variation

DNA copy number variation (CNV) has long been associated with specific chromosomal rearrangements and genomic disorders, but its ubiquity in mammalian genomes was not fully realized until 2006. Copy number variants account for a substantial amount of genetic variation. Since many CNVs include genes that result in differential levels of gene expression, CNVs may account for a significant proportion of normal phenotypic variation (Freeman *et al.*, 2006). A total of 1447 copy number variable regions, which can encompass overlapping or adjacent gains or losses, covering 360 megabases (12% of the human genome), were identified (Redon *et al.*, 2006). These sequences contained hundreds of genes, disease loci, functional elements, and segmental duplications. Notably, the copy number variable regions encompassed more nucleotide content per genome than SNPs, underscoring the importance of CNV in genetic diversity and evolution.

To date CNV has not been used significantly as a source of genetic polymorphism for detection or analysis of QTL. However, Maher (2008) proposed CNV as one of the reasons that only a small fraction of the total additive genetic variation in human height could be explained by genes detected in genome scans based on SNP.

Complete Genome Sequencing

The ultimate method for determining all variation in DNA is complete sequencing of the genome. The first DNA sequences were obtained in the early 1970s using laborious methods based on two-dimensional chromatography. Following the development of fluorescence-based sequencing methods with automated analysis, DNA sequencing became easier and orders of magnitude faster. Several new methods for high-throughput DNA sequencing were developed in the mid to late 1990s and were implemented in commercial DNA sequencers by the year 2000. In general these methods are termed "next-generation sequencing." Resequencing is necessary, because the genome of a single individual of a species will not indicate all of the genome variations among other individuals of the same species. All of these methods parallelize the sequencing process, producing thousands or millions of sequences concurrently. In ultrahigh-throughput sequencing as many as 500,000 sequencing-by-synthesis operations may be run in parallel.

These techniques have drastically lowered the cost of complete sequence of the genome. A $3-billion project to sequence the human genome was founded in 1990 by the US Department of Energy and the National Institutes of Health and was expected to take 15 years. A "rough draft" of the genome was finished in 2000. Ongoing sequencing led to the announcement of the essentially complete genome on

April 14, 2003, 2 years earlier than planned. By 2015 complete genome sequencing costs have been reduced to several thousand dollars per individual.

The 1000 Genomes Project was launched in January 2008 to sequence the genomes of at least one thousand anonymous participants from a number of different ethnic groups within 3 years. McVean *et al.* (2012) reported on the completion of the sequencing of 1092 human genomes. The complete genome of an individual cow was first sequenced in 2009 (Bovine Genome Sequencing and Analysis Consortium *et al.*, 2009). In 2012 the 1000 bull genomes project was initiated. Daetwyler *et al.* reported in 2014 on the complete sequencing of 234 bulls from different breeds to an average of 8.3-fold genome coverage.

Summary

In this chapter we reviewed the major milestones in the development of methodologies for high-throughput genotyping of large numbers of markers per individual. Since the original discovery of microsatellites in 1990, which were the first class of polymorphisms that made genome scans possible, costs were reduced from $10 per genotype to $0.002 per genotype. Among the SNP chips that are currently available for cattle are arrays that genotype 3000, 8000, 54,609, 139,480, 640,000, and 777,000 markers. Through 2015 genotyping costs have continued to decrease, making possible complete genome analyses based on next-generation sequencing methodologies at costs of several thousand dollars on the one hand and genotypes of several thousand markers at costs attractive to the individual farmer on the other.

3 Advanced Animal Breeding Programs Prior to Genomic Selection

Introduction

Before considering how marker-assisted selection (MAS) or genomic selection can be applied to animal breeding programs, it is necessary to understand the basic mechanics of animal breeding programs prior to MAS. All animal breeding programs are based on the principles of quantitative genetics, which will not be considered in details in this book. Advanced animal breeding programs can be divided into two groups: within-breed selection and programs based on crossbreeding among different breeds. Within-breed selection has been applied and studied most extensively for dairy cattle. Breeding programs based on crossbreeding are the norms for beef cattle, poultry, and swine. Crossbreeding programs can be further divided into those programs that are based on crossing two, three, or four breeds. The main advantages of crossbreeding schemes are twofold: utilization of heterosis and the fact that economic traits have different values in males and females. The disadvantage is the cost of maintaining the pure lines.

In the next section we will describe the basic principles used to evaluate selection within a breed. In the following section we will apply these principles to the specific problems related to dairy cattle breeding and the major breeding schemes that have been applied or proposed. In the following section we will also consider in more detail the advantages and limitations of crossbreeding programs, especially as related to MAS.

Within a Breed Selection: Basic Principles and Equations

The genetic gain due to selection within a breed per generation, Φ, will be a function of the selection intensity, i_s; the accuracy of the evaluation, ac; and the additive genetic standard deviation, σ_g. In most animal breeding schemes, the selection intensity and the accuracy of the evaluation will be different along the four paths of inheritance: sire to son, sire to daughter, dam to son, and dam to daughter. In general the genetic gain per generation along the four paths of inheritance can be computed by the following equation:

$$\Phi_i = i_{si}\text{ac}_i\sigma_g \tag{3.1}$$

where Φ_i is the genetic gain per generation for path i, i_{si} is the selection intensity for path i, ac_i is the accuracy of the genetic evaluation for path i, and σ_g is the genetic standard deviation, which will be

Genomic Selection in Animals, First Edition. Joel Ira Weller.

the same for all four paths of inheritance. The selection intensity is the difference between the mean of the individuals selected as parents and the general population mean in units of the standard normal distribution. If a fraction, p, of the population is selected to be parents of the next generation, then i_s can be computed as the density of the standard normal curve at the point of truncation divided by p (Falconer, 1964). The accuracy of the genetic evaluation is defined as the correlation between the genetic evaluation and the actual genetic value. Although the actual genetic value is unknown, the accuracy of the evaluation can be estimated as will be explained in Chapter 6, Section "Important Properties of Mixed Model Solutions." The square of the accuracy is termed the "reliability" of the evaluation. The annual gain for the entire population is then computed as

$$\Phi_y = \frac{\Phi_{ss} + \Phi_{sd} + \Phi_{ds} + \Phi_{dd}}{G_{ss} + G_{sd} + G_{ds} + G_{dd}} \tag{3.2}$$

where Φ_{ss}, Φ_{sd}, Φ_{ds}, and Φ_{dd} are the genetic gains per generation along the four paths of inheritance and G_{ss}, G_{sd}, G_{ds}, and G_{dd} are the generation intervals in years along the four paths.

Traditional Selection Schemes for Dairy Cattle

Dairy cattle are unique in that:

1. Males have nearly unlimited fertility via artificial insemination (AI), while females have very limited fertility.
2. Nearly all of the traits of interest are expressed only in females.

Since the mid-1980s it has become possible to increase fertility of females by multiple ovulation and embryo transplant, although these techniques are still relatively expensive.

Considering these limitations, most genetic gains are obtained by selection of males, even though the males can only be genetically evaluated based on the production records of their female relatives. Therefore commercial dairy cattle programs have traditionally been based on either half-sib or progeny test designs, described in detail by Owen (1975). Bulls reach sexual maturity of the age of 1 year. The male generation intervals in commercial breeding programs are usually much longer than the biological minimum. A typical half-sib breeding program is described in Figure 3.1, and a typical progeny test breeding program is described in Figure 3.2.

Both designs as described assume a total cow population of 100,000, but this is not a critical element of either design. Both designs have been applied to much larger populations. In the half-sib design, bull sires are selected based on the records of their daughters. These elite bulls are then mated to elite cows based on pedigree and their own production records. Of the 20 bull calves produced each year, about 10 are used for servicing the general cow population once they reach sexual maturity at the age of 1 year. Thus the bulls used for general service are selected based on the production records of the daughters of their sires, which are the half-sibs of the bulls used for general service. In this design the maximum accuracy of sire evaluations is 0.5, assuming that no information is available on the dam of the sire. With information on the dam, the accuracy can be slightly higher, but will not account for the "Mendelian sampling" of the two parental genotypes by the son.

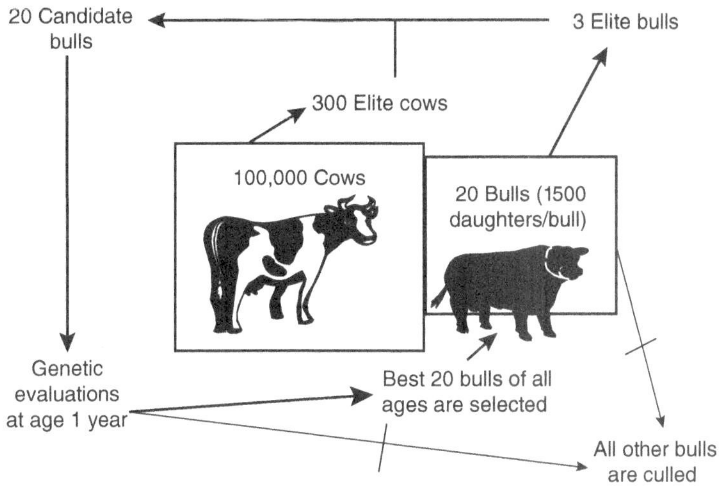

Figure 3.1 Typical half-sib test breeding program.

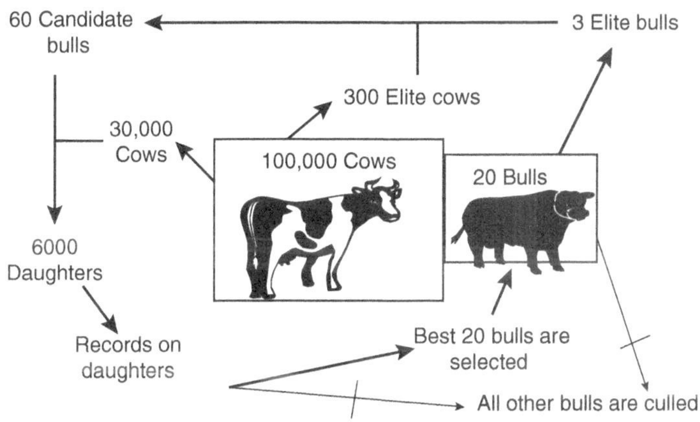

Figure 3.2 Typical progeny test breeding program.

Most advanced dairy cattle breeding programs are based on a progeny test of young sires based on a relative small sample of daughters. Sires with superior evaluations based on the first crop of daughters are returned to service. However, by the time daughter milk production records are available, these sires are 5 years old. As will be shown, theoretical studies demonstrate that the gain in accuracy obtained by the progeny test outweighs the loss incurred by increasing the generation interval.

In the progeny test design described in Figure 3.2, sires for general service are selected based on the production records of a sample of 50–100 daughters. Since the daughters completely reflect the additive genotype of the sire, it is possible with this design to approach an accuracy of unity for sire evaluations. With about 100 daughters, the accuracy of sire evaluations will be about 0.9. Thus the accuracy of the sire evaluations is nearly double by the progeny test scheme. Sires are used in general service only after their daughters complete their first lactation. As noted in the previous paragraph, by that time the sires are at least 5 years old.

Table 3.1 Expected annual genetic gains in units of the genetic standard deviation for the half-sib (HS) and progeny test (PT) designs for a trait with a heritability of 0.25.

Design	Path	Generation interval	Proportion selected	Selection intensity	Accuracy	Genetic gain[a]
HS	Sire to son	4.8	0.05	2.0	0.8	1.6
	Sire to daughter	2.5	1.00	0	0.6	0
	Dam to son	4.8	0.0017	3.2	0.7	2.24
	Dam to daughter	4.0	0.85	0.3	0.7	0.21
	Total	16.1				4.05
	Annual					**0.2516**
PT	Sire to son	7.4	0.02	2.4	0.95	2.28
	Sire to daughter (young)[b]	2.0	1.00	0	0.6	0
	Sire to daughter (proven)	7.4	0.11	1.7	0.95	1.614
	Dam to son	4.8	0.005	2.9	0.7	2.03
	Dam to daughter	4.0	0.85	0.3	0.7	0.21
	Total	22.5				5.796
	Annual					**0.2576**

[a] Computed as selection intensity multiplied by accuracy for each path.
[b] 21% of the cows are mated to young sires, and the remaining 79% are mated to proven sires.

The expected genetic gains in units of the genetic standard deviation by these two breeding schemes are summarized in Table 3.1, assuming that the breeding objective has a heritability of 0.25. Both schemes assume equal selection along the dam-to-daughter path. As noted earlier, selection intensity is low, because most female calves produced must be used as replacement cows. Although there is no selection along the sire-to-daughter path in the half-sib design, the expected annual genetic gain by this scheme is nearly equal to the genetic gain obtained by the progeny test design, because the mean generation interval is decreased.

Crossbreeding Schemes: Advantages and Disadvantages

"Heterosis" is generally defined as superiority of the hybrid over both parents (Strickberger, 1969). Moav (1966) defined five types of economic heterosis. That is, the economic value of the hybrid is greater than either parent. The main advantage of crossbreeding was termed by Moav (1966) "sire–dam" heterosis. Sire–dam heterosis is due to the fact that the values of the economic traits are different for males and females due to the more limited fertility of females in birds and mammals. Thus the economic value of traits related to female fertility in a "male" poultry broiler line is negligible, while the economic value of these traits can be of major importance in the "female" line. Therefore a cross between a male line with high growth rate but low female fertility and a female line with moderate growth rate but superior female fertility will result in greater economic value than either purebred line, even if the mean value of the progeny is at the biological mean for each of the individual traits.

In addition to "sire–dam" heterosis, heterosis is generally observed for the component traits included in profitability. Therefore to exploit heterosis, the parental lines are usually the result of crosses between four grandparental lines. Of course maintenance of the purebred lines is an extra cost that does not exist for selection programs based on selection within a single breed. Thus modern commercial poultry breeding programs are based on very large populations. Two companies currently control approximately 75% of the world broiler market. Key industry people agree that it takes a market share of 25–35% to enable basic breeders to make research investments of the size necessary to be competitive.

Selection for the component traits included in the selection objective is performed within the pure lines. Expected gains due to additive genetic variance can be predicted from the principles of quantitative genetics. However, this is not the case for heterosis. This complicates implementation of MAS and genomic selection and partially explains why to date there has been less adaption of genomic selection in crossbreeding programs. For a more extensive discussion of economic aspects of crossbreeding and heterosis, see Weller (1994).

Summary

Traditional selection index based on phenotypic records and information on relationships is very efficient, provided that it is possible to obtain high selection intensities, the selection criterion has a relative high heritability, and the selection criterion can be measured on all candidates for selection. However, many situations exist in which these conditions are not met. In many important mammalian species, such as cattle, the rate of genetic gain that can be obtained by traditional selection index methodology is limited, because the economic traits are expressed only in females, which have low fertility rates. Breeding schemes are based on genetic evaluations of males by their female relatives. It is in these situations that genomic selection can have a significant impact. With respect to crossbreeding programs, it is possible that genomic selection might help to limit some of the guesswork involved in exploiting heterosis.

4 Economic Evaluation of Genetic Breeding Programs

Introduction

In the previous chapter we described the basic methods of animal breeding programs prior to MAS and showed that advanced animal breeding programs are divided into two groups: within-breed selection and programs based on crossbreeding among different breeds. Several times in this chapter we referred to the "breeding objective" without defining the concept. Clearly the breeding objective should be considered in economic terms.

Generally genomic selection breeding programs have been evaluated merely in terms of the increase in accuracy of genomic evaluations, as compared to genetic evaluations based on phenotypic records and pedigree. An exception is Schaeffer (2006) who presented an economic evaluation, but only for the specific case of the Canadian dairy industry.

In economic evaluation of any enterprise, both returns and costs should be considered. Until the advent of genetic marker technology, costs of animal breeding programs were generally considered negligible with respect to the gains obtained, and optimization was generally considered only in terms of maximizing gain. With genomic selection this is no longer the case, and costs can be quite significant, especially if genotyping costs are borne by breeding companies or individual farmers. Thus it is necessary to decide what level of expenditure in marker technology can be economically justified. In the first section we will consider the nontrivial question of who is the "client" for breeding programs, and in the next section we will consider the criteria for economic evaluation of breeding programs.

National Economy versus Competition among Breeders

Economic evaluation of breeding programs must begin with the questions as to who is the "client" for breeding programs and who gains from genetic improvement. The economic entities involved in genetic improvement are farmers, food processing companies, breeding companies, governments, and the national economy. Although breeding companies generally try to convince farmers that the goal of animal breeding is to increase farmer profitability, this is rarely the case. In general, as shown by Moav (1973), due to competition among producers and breeding companies, the general public is the recipient of nearly all economic gain from breeding programs in the form of lower

Genomic Selection in Animals, First Edition. Joel Ira Weller.

prices or better products. Moav (1973) defined the "progress-surplus-bankruptcy cycle." If higher productivity is confined to a small group of farmers, then it will not affect the supply curve, and profits of farmers will increase. However, in nearly all cases, many producers will take advantage of genetic improvement. In a free market situation this will result in increased production and lowering of the market price. At the lower market price the cost of production for the least efficient producers will be above the market price, and they will go bankrupt. Other more efficient producers will be able to increase their market share.

However, in practice there is rarely a completely free market for agricultural produce, due to government intervention, which takes various forms. Less efficient producers can be kept in the market by price supports or imposition of production quotas, although these forms of intervention are becoming less common and often violate international trade agreements. A more politically acceptable form of intervention is government support for agricultural research and extension services. In practice the border between research and extension is often blurred. This is especially true for dairy cattle breeding, including genomics. In many countries the costs of genetic evaluation and genomic analysis are covered by government research institutions, even though routine running of genetic evaluation programs is not "research."

In some countries animal breeding, and especially dairy cattle breeding, is a quasigovernmental enterprise, and there is no competition within the country, although there generally is competition with breeding stock from foreign countries. In this case evaluation should be considered in terms of contribution to the national economy, that is, the effect of breeding on lowering the costs of production or increasing the quality of product.

In countries in which several breeding companies compete, evaluation of breeding programs should be considered in terms of increasing profit for the individual breeding companies. Although studies of this type have been performed, this type of analysis is problematic, because in most cases all competitors will incorporate more or less the same technology. Thus no one company will have an advantage over the other companies.

In addition to these two scenarios, it is also necessary to consider application of genomic selection by individual farmers. With the development of multiple ovulation and embryo transplant (MOET) in the 1980s, it became possible for the individual farmer to apply significant selection among dams within individual herds. Various studies performed economic evaluations of this and other fertility-related technologies to the individual farmer (e.g., Van Vleck, 1981, 1982). The general consensus was that application of MOET by the individual farmer could not be economically justified at the market costs during the 1980s. However, this is not the case for current genetic marker technologies, as will be seen in the following chapters.

Criteria for Economic Evaluation: Profit Horizon, Interest Rate, and Return on Investment

Considering the gain accrued to the national population, animal breeding programs differ from most other economic enterprises in three important aspects:

1. Due to biological limitations, especially the relatively long generation interval, animal breeding programs can only be evaluated over a long-term period of at least 10 years.
2. Genetic gains are generally cumulative.
3. Unlike other agri-technical gains, genetic gains are eternal. They do not "wear out," and no additional investment is required for maintenance.

Although genetic gains are permanent, the current value of these gains must be discounted as a function of time until the gains are achieved. Furthermore a breeding program is generally evaluated in terms of a profit horizon, under the assumption that gains obtained after the profit horizon have no current value. In order to simplify calculations the nominal value of genetic gains over the long term is generally considered to be a linear function of time, although, as shown by Weller (1994), this does not accurately reflect reality. Based on these assumptions, and assuming a lag time of several years until first gains are realized, Hill (1971) presented the following equation for the cumulative discounted returns of a breeding program, R:

$$R = V\left[\frac{r^t - r^{T+1}}{(1-r)^2}\right] - \left[\frac{(T-t+1)r^{T+1}}{1-r}\right] \tag{4.1}$$

where

V = the nominal value of 1 year of genetic gain
$r = 1/(1+d)$, where d = the yearly discount rate
T = profit horizon in years
t = the lag time until first returns are realized in years

To illustrate the huge value of genetic gain as compared to other investments, assume a discount rate of 8%, a lag of 5 years, and a profit horizon of 20 years. In this case, $R = 32.58V$!

Unlike genetic gain, costs of a breeding program are not cumulative, but begin immediately without any lag. Assuming equal costs for each year of the breeding program, cumulative costs, C, can be computed as follows (Hill, 1971):

$$C = \frac{C_c r(1-r^T)}{1-r} \tag{4.2}$$

where C_c is thethe annual costs of the breeding program, and the other terms are as described previously. Using the same values for r and T gives $C = 9.82C_c$. Thus with these values, a breeding program will be profitable within 20 years even though annual costs are threefold the nominal value of the annual genetic gain.

Weller (1994) considered in detail the appropriate criteria for economic evaluation of breeding program. Four alternatives were proposed:

1. Assume that the discount rate and profit horizon are fixed, and compute aggregate profit until the profit horizon is reached.
2. Assume that the profit horizon is fixed, and estimate the discount rate necessary to achieve a net profit of zero at the profit horizon.
3. Assume the discount rate is fixed, and estimate the number of years required to achieve a net profit of zero.
4. Assume a fixed discount rate and a profit horizon of infinity, but compute profit from only a single cycle of selection.

Weller (1994) presented calculations to optimize investment in breeding programs, although these equations require major simplifications of any true breeding program and have not been applied in practice.

Summary

In this chapter we considered the basic principles of economic evaluation of breeding programs, noting the important economic differences between these programs and nearly all other types of economic enterprises. Although various studies have attempted to calculate economic evaluation of MAS programs under situations of competition among a number of breeding companies, we showed that in most cases economic evaluation of breeding programs should be made in terms of their contribution to the national economy. Very few studies have actually attempted to economically optimize breeding programs. In most cases only two scenarios were compared. In addition there is not a clear consensus as to the appropriate criterion for economic evaluation of breeding programs, and four different criteria considered in the literature were presented.

5 Least Squares, Maximum Likelihood, and Bayesian Parameter Estimation

Introduction

Variables are generally divided into two groups, fixed and random. Random variables are assumed to be sampled from a distribution with known parameters, while no such assumptions are made about fixed variables. Fixed variables are also denoted "parameters." The most common method of parameter estimation is least squares estimation (LSE), which is based on deriving the parameter estimates that minimize the expectation of the sum of squared errors. Thus, by definition this method has minimum estimation error variance. LSE is described in detail in many statistic texts and will therefore be described only briefly in this book. We will consider in detail maximum likelihood (ML) and Bayesian estimation, because of their relevance to genomic selection, and the fact that they are generally not considered in detail in basic statistic courses.

A basic understanding of matrix algebra is required to understand the remainder of this text. See Searle (1982) for an extensive study of matrix algebra as applied to statistics. For a more abbreviated summary of topics required in this text, see Weller (1994). Throughout this text, we will use the conventions of denoting matrices in upper case **bold type**, vectors in lower case **bold type**, and scalar variables in *italics*. The identity matrix will be denoted **I**, a transpose of a matrix by an apostrophe, and the inverse of a matrix by the −1 power.

Least Squares Parameter Estimation

LSE is based on deriving the parameter estimates that minimize the expectation of the sum of squared errors. Thus, by definition this method has minimum estimation error variance. In matrix form a completely general model can be written as follows:

$$\mathbf{y} = f(\theta') + \mathbf{e} \tag{5.1}$$

where **y** is the vector of observations, θ is the vector of parameters, $f(\theta')$ is some function of θ, and **e** is the vector of residuals. The least squares solution, $\hat{\boldsymbol{\theta}}$, is the vector that minimizes $[\mathbf{y} - f(\hat{\boldsymbol{\theta}})]^2 = \mathbf{e}^2$. For a linear model, this equation can be written as follows:

$$\mathbf{y} = \mathbf{X}\boldsymbol{\theta} + \mathbf{e} \tag{5.2}$$

Genomic Selection in Animals, First Edition. Joel Ira Weller.

where **X** is a matrix of coefficients of θ. Effects in linear models can take one of two forms—class or continuous. Discrete effects such as a specific herd, block, or sex are denoted "class effects." Although the levels of these effects can be numbered, there is no relationship between the number of a specific herd and effect associated with it. For continuous effects a linear relationship is assumed between the value for the independent variable and the dependent variable. Each row of **X** corresponds to the coefficients of θ for a specific record in **y**. For class effects the elements in **X** will be either zero or one. For continuous effects, each element in **X** corresponds to the observed value for the independent variable.

The least squares solutions are solved by finding the parameter estimates that minimize the sum of squares of the residuals.

The residual sum of squares in matrix notation is computed as follows:

$$(\mathbf{y}-\mathbf{X}\boldsymbol{\theta})'(\mathbf{y}-\mathbf{X}\boldsymbol{\theta})=\mathbf{e}'\mathbf{e} \tag{5.3}$$

$$\mathbf{y}'\mathbf{y}-2(\mathbf{X}\boldsymbol{\theta})'\mathbf{y}+(\mathbf{X}\boldsymbol{\theta})'\mathbf{X}\boldsymbol{\theta}=\mathbf{e}'\mathbf{e} \tag{5.4}$$

Setting the differential with respect to θ equal to zero and solving give

$$\boldsymbol{\theta}=(\mathbf{X}'\mathbf{X})^{-1}\mathbf{X}'\mathbf{y} \tag{5.5}$$

These equations are termed the "normal equations" and are used extensively in modern statistics. If the observations are correlated or do not have equal variances or both, then the normal equations can be modified as follows:

$$\mathbf{X}'\mathbf{V}^{-1}\mathbf{X}\boldsymbol{\theta}=\mathbf{X}'\mathbf{V}^{-1}\mathbf{y} \tag{5.6}$$

where **V** is the variance matrix among the observations. **V** is a diagonal matrix with rows and columns equal to the number of observations. The diagonal elements of **V** are the variance of each observation, and the off-diagonal elements are the covariances between the corresponding pair of observations. Solutions to these equations are called "generalized least squares" solutions, and minimize **e′e**, subject to the restriction of the known variance matrix. Solution of these equations requires the inverse of **V**, which is difficult to compute for large data sets.

If **y** is not a linear function of θ, then the least squares solution can generally not be derived analytically, although various iterative methods have been developed. Only effects on the mean of **y** are included in the model; thus effects on the variance of **y** or higher-order moments cannot be estimated by least squares.

ML Estimation for a Single Parameter

ML is much more flexible than LSE but generally requires rather complex programming. There are three steps in ML parameter estimation:

1. Defining the assumptions on which the statistical model is based.
2. Constructing the likelihood function; this is the joint density of the observations conditional on the parameters.
3. Maximizing the likelihood function with respect to the parameters.

The basic methodology for ML estimation of a single parameter will be illustrated using an example from a binomial distribution. Assume that from a sample of 10 observations, 3 are "successes" and the other 7 are "failures." We wish to derive the ML estimate (MLE) of p, the probability of success. The binomial probability for this result as a function of p is

$$L = \frac{10!\,p^3(1-p)^7}{3!7!} \tag{5.7}$$

where L is the probability of obtaining this result, conditional on p. L is denoted the "likelihood function." The MLE for p is that value of p which maximizes L. The MLE is computed by differentiating L with respect to p and solving for p, with this derivative set equal to zero. In practice it is usually easier to compute and differentiate the log of L. With respect to ML, this is equivalent to differentiating L, because a function of a variable and the log of the function will be maximal for the same value of the variable. The MLE of p is then derived as follows:

$$\text{Log}\,L = \text{Log}(10!) - \text{Log}(3!7!) + 3(\text{Log}\,p) + 7[\text{Log}(1-p)] \tag{5.8}$$

$$\frac{d(\text{Log}\,L)}{dp} = \frac{3}{p} - \frac{7}{1-p} = 0 \tag{5.9}$$

$$p = \frac{3}{10} \tag{5.10}$$

This is, of course, the proportion of successes derived in the sample. Thus, for this simple case, the MLE is the intuitive estimate value. From the earlier discussion, it should be clear why MLE must lie within the parameter space. A parameter estimate outside the parameter space will, by definition, have a likelihood of zero and can therefore not be the MLE.

For a continuous distribution, the likelihood is computed as the statistical density of the distribution, conditional on the sample. Statistical density, $f(y)$, for a continuous variable, y, is defined as the ordinate of the distribution function for a given value of y. For example, assume that a sample was taken from a normal distribution. To obtain the MLE for the mean, it is necessary to compute the joint statistical density of the sample. For a single observation the likelihood will be

$$L = \frac{e^{-(y-\mu)^2/2\sigma^2}}{\sqrt{2\pi\sigma^2}} \tag{5.11}$$

where σ is the standard deviation, e is the base for natural logarithms and is approximately equal to 2.72, μ is the mean, π is the ratio of the circumference and the diameter of a circle (~3.141), and y is the variable value. For a sample of n observations, the likelihood will be the product of the likelihoods for each individual observation. As in the previous case, the MLE for μ can be derived by computing the derivative of the log of the likelihood with respect to the mean and setting this function equal to zero. The derivative of Log L for a sample from a normal distribution is computed as follows:

$$L = \prod^{I}\left[\frac{e^{-(y_i-\mu)^2/2\sigma^2}}{\sqrt{2\pi\sigma^2}}\right] \tag{5.12}$$

$$\text{Log } L = \sum^{I} \left[\frac{e^{-(y_i-\mu)^2/2\sigma^2}}{\sqrt{2\pi\sigma^2}} \right] \tag{5.13}$$

$$\frac{d(\text{Log } L)}{d\mu} = \Sigma(y_i - \mu) \tag{5.14}$$

where Π signifies a multiplicative series, parallel to Σ, and y_i is element i of y. Setting $\Sigma(y_i-\mu)$ equal to zero, we find that the MLE of μ is $(\Sigma y_i)/n$, the sample mean, which is again the intuitively correct result.

The MLE for the variance could be derived in the same manner and would again yield the intuitive result of the sample variance, that is, $[\Sigma(y_i-\mu)^2]/n$. Note that if the objective is to derive an estimate of the variance for the population, and not the sample, then division should be by $n-1$, instead of n. Thus the estimate $[\Sigma(y_i-\mu)^2]/n$ will be a biased sample of the population variance. This problem can be solved by application of "restricted maximum likelihood estimation," generally denoted REML. For a detailed explanation of REML, see Lynch and Walsh (1998).

Although in the two examples given so far, ML has been used to derive estimates that could have been derived by other methods, for more complicated problems, ML estimation and Bayesian estimation are the only estimation methods that can utilize all the available data.

ML Multiparameter Estimation

ML can also be used to estimate several parameters simultaneously, for example, to estimate both the mean and variance in a normal distribution. In that case it is necessary to maximize the likelihood with respect to both parameters. This can be done by computing the partial derivatives of the log-likelihood with respect to each parameter and setting each partial derivative equal to zero. It is then necessary to solve a system of equations equal to the number of parameters being estimated. In general the likelihood function for the estimation of m parameters, $(\theta_1, \theta_2, \ldots, \theta_m)$, from a sample of n observations $(y_1, y_2, \ldots, y_n)$ can be written as follows:

$$\begin{aligned} L &= p(y_1, y_2, \ldots, y_n \mid \theta_1, \theta_2, \ldots, \theta_m) \\ &= p(y_1 \mid \theta_1, \theta_2, \ldots, \theta_m)\, p(y_2 \mid \theta_1, \theta_2, \ldots, \theta_m) \cdots p(y_n \mid \theta_1, \theta_2, \ldots, \theta_m) \\ &= \Pi p(y_i \mid \theta_1, \theta_2, \ldots, \theta_m) \\ &= \Pi p(y_i \mid \boldsymbol{\theta}) \end{aligned} \tag{5.15}$$

where $p(y_i|\theta)$ represents the probability of obtaining y_i, conditional on the vector of parameters. If the distribution is continuous, then $p(y_i|\theta)$ will be replaced by $f(y_i|\theta)$, that is, the density of y_i, conditional on θ. Thus, ML can be applied to solve any problem that can be phrased in terms of this equation.

As an example we will consider the relatively simple case of ML estimation of the parameters for a QTL effect as estimated from a backcross (BC) design with a single genetic marker. This design is diagrammed in Figure 5.1. Two parental strains differing in both marker and QTL genotypes are mated to produce an F-1. It is generally assumed that the two parental strains are homozygous for alternate alleles of both loci. Thus all F-1 individuals will have the same heterozygous genotype.

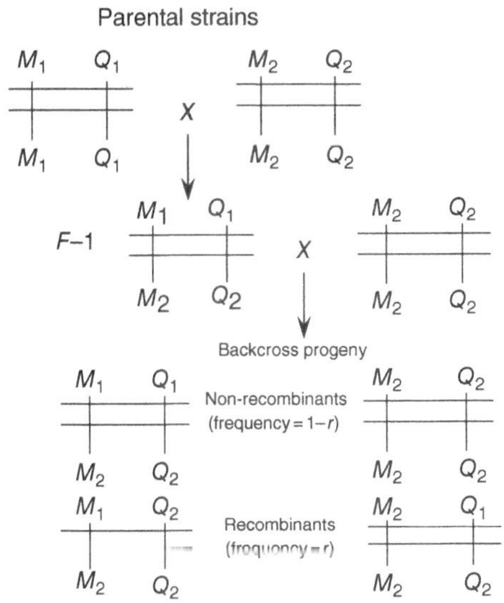

Figure 5.1 The backcross design.

The F-1 is then mated to one of the parental strains. The genetic background for this cross is then three-quarters of the recurrent parent and one-quarter of the other parent. The BC progeny is divided into two groups, based on their marker genotypes. All loci not linked to the genetic marker under consideration should be randomly distributed among the marker genotype groups. With a single marker there are only two marker genotype groups for the BC design. We will assume that the residuals are normally distributed with equal variances. The statistical density function for a single individual of genotype M_1M_2 will be

$$L = \frac{(1-r)e^{-(y-\mu_1)^2/2\sigma^2}}{\sqrt{2\pi\sigma^2}} + \frac{(r)e^{-(y-\mu_2)^2/2\sigma^2}}{\sqrt{2\pi\sigma^2}} \tag{5.16}$$

where y is the trait value, σ is the standard deviation, μ_1 is the mean of individuals with the Q_1Q_2 genotype, μ_2 is the mean of individuals with the Q_2Q_2 genotype, and r is the recombination frequency between the marker and the QTL. Individuals with the M_2M_2 genotype will have the same likelihood, except that the QTL mean values will be reversed. The complete likelihood for a sample of individuals can be written as follows:

$$L = \prod^{N_1}\left[f(y_i, M_1M_2)\right]\prod^{N_2}\left[f(y_j, M_2M_2)\right] \tag{5.17}$$

where $f(y_i, M_1M_2)$ and $f(y_j, M_2M_2)$ are the statistical densities for ith and jth observations with genotypes M_1M_2 and M_2M_2, respectively, and N_1 and N_2 are the number of individuals with the two genotypes, respectively.

To obtain the ML parameter estimates, the log of this function must be differentiated with respect to four parameters: μ_1, μ_2, σ, and r. The partial derivatives must then be equated to zero, and this system of four equations must be solved.

Although it is generally possible to write the likelihood function and differentiate Log L with respect to the different parameters, even for the relatively simple example given, it will not be possible to solve analytically the resultant system of equations. Iterative methods to derive solutions have been developed and will be considered in the following section.

Methods to Maximize Likelihood Functions

Numerous iterative methods have been proposed to maximize multiparameter likelihood function. Generally the initial solutions for all methods are selected arbitrarily or set to zero. Of all the methods proposed, only expectation–maximization (EM) is guaranteed to converge to a maximum, provided a maximum exists within the parameter space. However, even for EM, the convergence point may be only a local maximum.

Iterative maximization methods can be divided into three categories: derivative-free methods, methods based on computation of first derivatives, and methods based on computation of second derivatives. For all derivative-based methods, the parameter estimates of the ith iterate are computed by solving a system of equations equal in number to the number of parameters being estimated. These reduced equations are themselves functions of the parameter estimates from the previous iteration. Generally, iteration is continued until changes between rounds fall below a sufficiently small value.

Although this is the generally accepted criterion for approximate convergence, this is not necessarily the case. If the convergence is slow, it is possible that changes between consecutive rounds of iteration can be small, even if the estimates are not close to the actual solutions. Convergence is generally most rapid for second derivative methods, but the convergence is not guaranteed, even if there is a maximum within the parameter space.

Confidence Intervals and Hypothesis Testing for MLE

In addition to deriving parameter estimates, it is also important to determine the accuracy of the estimates. Generally the standard errors of the estimates are used for this purpose. The square of the standard error is denoted the "prediction error variance." The following equation can generally be used to derive the prediction error variance (PEV) for MLE of a single parameter:

$$\mathrm{PEV}\left(\hat{\theta}\right) = \frac{-1}{E\left[d^2\left(\mathrm{Log}\,L\right)/d\theta^2\right]} \tag{5.18}$$

where $\hat{\theta}$ is the MLE of θ, and $E[d^2(\mathrm{Log}\,L)/d\theta^2]$ is the expectation of the second derivative of L with respect to θ. This equation will be correct if the first derivative of θ is a multiple of the difference between the true parameter value and its estimate. Otherwise the prediction error variance will be slightly greater than the right-hand side of this equation. Under a wide range of conditions, this equation will be "asymptotically correct"; that is, as the sample size increases, the difference between the left-hand and right-hand sides of the equation tends toward zero. The square root of the prediction error variance, the standard error of the estimate, can be used to determine the confidence interval of the estimate.

The prediction error variances for the multiparameter estimation problem can be derived in a manner parallel to that described in the previous equation. The parameter estimates and the first derivatives will each consist of a vector with the number of elements equal to m, the number of parameters. The second derivatives and the prediction error variances will both be square $m \times m$ matrices. Using brackets to denote matrices and vectors, the matrix of prediction error variances can be computed with the following equations:

$$\mathrm{PEV}\left(\hat{\theta}\right) = \left[\frac{\partial^2 \mathrm{Log}\, L}{\partial [\theta]^2}\right]^{-1} \tag{5.19}$$

where the right-hand side of the equation is the inverse of the matrix of second partial derivatives with respect to $[\theta]$. The diagonal elements will be the prediction error variances of the estimates, and the off-diagonal elements will be the prediction error covariances between the elements. These are needed to test hypotheses based on linear functions of the parameters.

Even if the prediction error variance is not computed, ML can still be used to test a hypothesis by a "likelihood ratio test." In a likelihood ratio test the ML obtained under two alternative hypotheses are compared. In the null hypothesis, one or more of the parameters that are maximized in the alternative hypothesis are assumed fixed. For example, the mean is set equal to zero. The alternative hypothesis is termed the "complete" model, because MLE are derived for all parameters, while the null hypothesis is termed the "reduced" model, because some of the parameter values are fixed. Under the assumption that the null hypothesis is correct, the natural log of the ML ratio of the complete and reduced models will be asymptotically distributed as $(1/2)\chi^2$, where χ^2 is the chi-squared statistic. The number of degrees of freedom (dof) will be equal to the number of parameters that are maximized in the alternative hypothesis but fixed in the null hypothesis. This ratio will have a χ^2 distribution only if the null hypothesis is "nested" within the alternative hypothesis. Hypothesis is "nested" if some parameters that are fixed in the null hypothesis are set to their ML values in the alternative hypothesis, but all parameters that are fixed in the alternative hypothesis are also fixed in the null hypothesis.

Bayesian Estimation

Bayesian estimation differs from ML in that instead of maximizing the likelihood function, the "posterior probability" of θ, $p(\theta|y)$, is maximized as a function of the likelihood function multiplied by the "prior" distribution of θ. Bayes' theorem in general terms for multiple parameters and observations can be written as follows:

$$p(\theta_1, \theta_2, \ldots, \theta_m \mid y_1, y_2, \ldots, y_n) = p(\theta_1, \theta_2, \ldots, \theta_m)\, p(y_1, y_2, \ldots, y_n \mid \theta_1, \theta_2, \ldots, \theta_m) \tag{5.20}$$

where $p(\theta_1, \theta_2, \ldots, \theta_m|y_1, y_2, \ldots, y_n)$ is the "posterior" probability of the parameters, $p(\theta_1, \theta_2, \ldots, \theta_m)$ is the "prior probability" of the parameters, and $p(y_1, y_2, \ldots, y_n|\theta_1, \theta_2, \ldots, \theta_m)$ is the likelihood function. Similar to ML, it is possible to maximize the posterior probability or density function relative to the parameter values. Assuming that prior information of the parameters is available, Bayesian estimation, which makes use of this information, should be preferable to ML, which ignores any prior information on the parameters.

Instead of maximizing the posterior density, it is possible to define a "loss function" which determines the economic value "lost" by incorrect parameter estimation. Common examples are linear and quadratic loss functions. In the linear loss function, the value of the loss is a linear function of the difference between the parameter estimates and their true values. In the quadratic loss function, the loss increases quadratically as a function of the difference between the parameter estimate and its true value. Minimizing the linear loss function is equivalent to maximizing the posterior density. Minimizing the quadratic loss function is equivalent to maximizing the mean of the posterior distribution.

Similarly a Bayesian test of alternative hypothesis is based on minimizing the expectation of the loss function. If a decision must be made between two alternative hypotheses, the economic value of the "loss" is determined for each incorrect decision. The expectation of the loss will be the probability of each incorrect decision (the type I and type II errors) multiplied by its economic value. The decision is then based on minimizing the expected loss.

There are two major drawbacks to Bayesian estimation. First, prior information on the parameters is often vague, and it is not possible to mathematically represent this information in terms of a statistical distribution function without additional assumptions, which cannot be verified. Second, if many records are included in the analysis, then the likelihood function tends to "overwhelm" the prior distribution of θ. In this case, the Bayesian estimates tend to converge to the MLE.

Parameter Estimation via the Gibbs Sampler

The Gibbs sampler is an algorithm to generate a sequence of samples from the joint probability distribution of two or more random variables. The purpose of such a sequence is to approximate the joint distribution and to approximate the marginal distribution of one of the variables, or some subset of the variables, such as the unknown parameters. Gibbs sampling is applicable when the joint distribution is not known explicitly or is difficult to sample from directly, but the conditional distribution of each variable is known and is easy (or at least easier) to sample. Gibbs sampling is particularly well adapted to sampling the Bayesian posterior distribution, which is typically specified as a collection of conditional distributions.

The advantage of Gibbs sampling is that given a multivariate distribution, it is simpler to sample from a conditional distribution than to marginalize by integrating over a joint distribution. Suppose we want to obtain k samples of $\boldsymbol{x}=\{x_1, \ldots, x_n\}$ from a joint distribution $p(x_1, \ldots, x_n)$. Denote the ith sample by $\boldsymbol{x}^{(i)}=\{x_1^{(i)}, \ldots, x_n^{(i)}\}$. We proceed as follows:

1. We begin with some initial value $\boldsymbol{x}^{(0)}$ for each variable.
2. For each sample $i=\{1, \ldots, k\}$, sample each variable $x_j^{(i)}$ from the conditional distribution $p(x_j^{(i)}| x_1^{(i)}, \ldots, x_{j-1}^{(i)}, x_{j+1}^{(i-1)}, \ldots, x_n^{(i-1)})$. That is, sample each variable from the distribution of that variable conditioned on all other variables, making use of the most recent values and updating the variable with its new value as soon as it has been sampled.

The samples then approximate the joint distribution of all variables. Furthermore, the marginal distribution of any subset of variables can be approximated by simply examining the samples for that subset of variables, ignoring the rest. In addition, the expected value of any variable can be approximated by averaging over all the samples. An example application of the Gibbs sampler will be given in section "Estimation of Variance Components via the Gibbs Sampler" of Chapter 8.

Generally, the samples at the beginning are discarded (the so-called burn-in period), and then only every nth sample is used when averaging values to compute an expectation. For example, the first 1000 samples might be ignored, and then every 100th sample is retained, discarding all the rest. The reason for this is that (i) successive samples are not independent of each other and (ii) the stationary distribution is the desired joint distribution over the variables, but it may take awhile for that stationary distribution to be reached.

Summary

In this chapter we describe in general terms the main statistical methodologies used for estimation of parameters, or fixed variables. Although least squares is used almost exclusively to estimate parameters, ML and Bayesian estimation methods are used to estimate both fixed and random variables. ML is much more flexible than LSE and guarantees that the estimates are within the parameter space. However, for models of interest, solutions can only be derived by iterative methods.

Bayesian methods are an extension of ML but differ from ML in that they require determination of the prior distribution of the parameters. This information is often lacking or vague. Of the methods considered in this chapter chiefly Bayesian methods have been used for genomic evaluations, and these methods will be considered in more detail in Chapters 15, 16, and 17.

In the following chapter we will consider methods for estimation of random variables in more detail, especially models that include both fixed and random variables, termed "mixed models." The mixed model is the base from which nearly all methods of genomic evaluation are derived.

6 Trait-Based Genetic Evaluation: The Mixed Model

Introduction

Hazel in 1943 formulated the principles of economic selection index. He asked the following question: "Assume that there are n traits for which breeding values can be estimated, and m traits with economic values. Assume further that the economic values of the m traits are linear functions of the trait values. What linear index of the n measured traits should be used to select individuals so as to maximize genetic progress on the economic scale?"

As noted in the previous chapter, random effects differ from fixed effects in that it is assumed that each observed random effect is sampled from an infinite population of possible effects with a known distribution. The basis of selection index theory is that polygenic breeding values for quantitative traits should be considered random, because these effects are "sampled" from a normal distribution of genetic values with a specific variance.

In genetic evaluation based on field records, it will also be necessary to include other effects, such as herd or block, in the model. These "nuisance" effects will generally be considered fixed. (The reasons for this have been discussed in detail by many sources (e.g., Henderson, 1984) and are beyond the scope of this text.) Therefore, analysis models will include both fixed and random effects in addition to the residual. The models that include both fixed and random effects are termed "mixed models."

We will first consider selection index theory, assuming the absence of fixed "nuisance" effects. We will then consider the general strategy for solving mixed models, based on the "mixed model equations" (Henderson, 1973). In general it will be assumed that random effects are sampled from a normal distribution with a mean of zero and a known variance. Therefore, estimates for random effects can only be derived if their variances are known. In the final sections we will consider methods for variance component estimation in mixed models, based on constant fitting and maximum likelihood and restricted maximum likelihood (REML).

Principles of Selection Index

Generally several traits have economic values in a species under selection. How then should selection be performed so as to economically maximize genetic improvement? We will start by assuming that for each individual there is a vector **y**, of length m, consisting of the individual's breeding values for traits of economic importance and a vector **x** of n measured traits to be included in the selection

Genomic Selection in Animals, First Edition. Joel Ira Weller.

index. Although **x** and **y** may include the same traits, this does not have to be the case. Assume further that the "economic values" associated with **y** are linear functions of the trait values. We can then define a vector **a**, also of length *m*, consisting of the economic values of the traits in **y**. The aggregate economic breeding value, *H*, can then be computed as **a′y**. The units of **y** are trait units, and the units of **a** are monetary units/trait units, for example, dollars/kilogram milk. Thus *H* is a scalar in monetary units. *H* is the "optimum" selection index. By this we mean that for a given selection intensity, the response to selection will be greatest, in monetary units, if candidates for selection are ranked by *H*.

Since the elements of **y** are generally unknown, the goal is to derive the linear index, I_s, of **x**, which is most similar to *H*. By "most similar" we mean either to maximize the correlation or to minimize the mean squared deviation between I_s and *H*. Specifically, if **b** is defined as a vector of index coefficients, then $I_s = \mathbf{b}'\mathbf{x}$, and the objective is to solve for **b** that maximizes the correlation between **b′x** and **a′y**. Of course, like *H*, I_s will be a scalar in monetary units.

To derive I_s we will define three additional matrices: **P**, the $n \times n$ phenotypic variance matrix of the traits in **x**; **C**, the $n \times m$ genetic covariance matrix between the measured traits in **x** and the breeding values in **y**; and **G**, the $m \times m$ genetic variance matrix for the traits in **y**. The selection index coefficients are then derived from the following equation:

$$\mathbf{b} = \mathbf{P}^{-1}\mathbf{C}\mathbf{a} \tag{6.1}$$

Brascamp (1984) presents several methods to derive this equation. We will present only one method, based on minimizing the squared difference between I_s and *H*. This is also equivalent to maximizing the correlation between I_s and *H* and maximizing the expected mean breeding value of individuals selected based on I_s. The derivation is simplified by assuming that both **x** and **y** are measured relative to their means. It is then necessary to minimize the following function:

$$(I_s - H)^2 = (\mathbf{b}'\mathbf{x} - \mathbf{a}'\mathbf{y})^2 \tag{6.2}$$

The expectation of the left-hand side of this equation can be computed as follows:

$$E(\mathbf{b}'\mathbf{x} - \mathbf{a}'\mathbf{y})^2 = E(\mathbf{b}'\mathbf{x}\mathbf{x}'\mathbf{b} - 2\mathbf{b}'\mathbf{x}\mathbf{y}'\mathbf{a} + \mathbf{a}'\mathbf{y}\mathbf{y}'\mathbf{a}) \tag{6.3}$$

Since **x** and **y** are scored relative to their means, **xx′** and **yy′** will be the variance matrices for **x** and **y**, and **xy′** will be the covariance matrix between them. Thus

$$E(\mathbf{b}'\mathbf{x} - \mathbf{a}'\mathbf{y})^2 = \mathbf{b}'\mathbf{P}\mathbf{b} - 2\mathbf{b}'\mathbf{C}\mathbf{a} + \mathbf{a}'\mathbf{G}\mathbf{a} \tag{6.4}$$

with all terms as defined earlier. Differentiating with respect to **b** and equating to zero, we obtain

$$2\mathbf{P}\mathbf{b} - 2\mathbf{C}\mathbf{a} = 0 \tag{6.5}$$

$$\mathbf{P}\mathbf{b} = \mathbf{C}\mathbf{a} \tag{6.6}$$

Solving for **b**, we obtain the selection index equations. If all traits included in the aggregate genotype are also included in the index, then **G** = **C**, and

$$\mathbf{b} = \mathbf{P}^{-1}\mathbf{G}\mathbf{a} \tag{6.7}$$

This is the selection index equation most commonly presented.

Of all possible linear indices of **x**, the selection index will have the highest correlation with H and the lowest squared deviation. In addition, selection of individuals on I_s will result in maximum expected mean value for H of the selected individuals, and genetic response to selection on I_s will be greater than for selection on any other linear index of **x**. These and a few other properties of the selection index are summarized by Henderson (1973). We will now describe some additional useful properties of selection index, based on Cunningham (1969), James (1982), and Lin (1978).

From the previous derivation, it should already be clear that the variance of the selection index can be computed as follows:

$$\sigma_{Is}^2 = \mathbf{b'Pb} = \mathbf{a'C'P^{-1}Ca} \tag{6.8}$$

The variance of the aggregate breeding value will be $\mathbf{a'Ga}$. The covariance between I and H can be computed as follows:

$$\sigma_{H,Is} = \mathbf{a'yx'b} = \mathbf{a'Cb} = \mathbf{a'CP^{-1}Ca} = \sigma_{Is}^2 \tag{6.9}$$

That is, the variance of the index is also equal to the covariance between I_s and H. Since this is the case, the correlation between H and I_s, r_{HI}, will be equal to $\left[\sigma_{Is}^2/\sigma_H^2\right]^{0.5}$. This correlation for the selection index is parallel to the "accuracy" of single-trait genetic evaluation, given in Equation (3.1). Thus the response to selection on the index, Φ_I, can be computed as follows:

$$\Phi_I = i_s r_{HI}\sigma_H = i\sigma_{Is} \tag{6.10}$$

where i_s is the selection intensity and σ_H and σ_{Is} are the standard deviations of H and I_s, respectively. As noted in Chapter 3, the selection intensity is the difference between the mean of the individuals selected as parents and the general population mean in units of the standard normal distribution. Φ_I will also be measured in monetary units. Thus the response to selection will be a direct function of the selection intensity and the standard deviation of the index.

Finally it is often of interest to compute the expected responses of the component traits to selection on the index. The genetic change for the ith trait due to selection on the index, ϕ_i, is computed as follows:

$$\phi_i = i_s b_{giI}\sigma_{Is} = i_s\left[\frac{\mathrm{Cov}(g_i, I_s)}{\sigma_{Is}^2}\right]\sigma_{Is} = \frac{i_s\left[\mathrm{Cov}(g_i, I_s)\right]}{\sigma_{Is}} \tag{6.11}$$

where b_{giI} is the genetic regression of the ith trait on the index and $\mathrm{Cov}(g_i,I_s)$ is the covariance between the genetic value of the ith trait and the index. $\mathrm{Cov}(g_i,I_s)=\mathrm{Cov}(g_i, \mathbf{p'b})=[\mathrm{Cov}(g_i, \mathbf{p'})]\mathbf{b}$, where $\mathrm{Cov}(g_i, \mathbf{p'})$ is the ith column of **C**. Thus the vector of correlated responses for all traits, ϕ, is computed as follows:

$$\phi = \frac{i_s\mathbf{Cb}}{\sigma_{Is}} \tag{6.12}$$

If all traits included in H are included in the index, then **C** can be replaced with **G**.

The Mixed Linear Model

As noted by Henderson (1973), selection index can be used to determine breeding values of animals, provided the population mean is known and no other effects bias the calculation. With field data this is rarely the case. The sample mean and the effects of other factors, such as herd or block, must be estimated from the data. These "nuisance" effects will generally be considered fixed effects, as opposed to genetic effects, which will be considered random.

As an example we will first consider the following simple mixed model used to derive breeding values of bulls for milk production:

$$Y_{ijk} = H_i + S_j + e_{ijk} \tag{6.13}$$

where Y_{ijk} is the milk production record of cow k in herd i, H_i is the effect of herd i, S_j is the effect of the cow's sire j on her production, and e_{ijk} is the random residual. The herd effect will be assumed to be a fixed effect, and the sire effect will be assumed to be random. In general terms the mixed model can be written in matrix notation as follows:

$$\mathbf{y} = \mathbf{X}\boldsymbol{\beta} + \mathbf{Zu} + \mathbf{e} \tag{6.14}$$

where $\boldsymbol{\beta}$ is a vector of fixed effects, $\mathbf{u}$ represents the vector of random effects, $\mathbf{X}$ and $\mathbf{Z}$ are incident matrices, and $\mathbf{e}$ is the vector of random residuals. The additive breeding values are considered random effects, with a known variance matrix. Both $\mathbf{u}$ and $\mathbf{e}$ are assumed to have a normal distribution. Thus $\mathbf{y}$ has a multivariate normal distribution with a mean of $\mathbf{X}\boldsymbol{\beta}$ and a variance $\mathbf{V}$ computed as follows:

$$\mathbf{V} = \mathbf{ZGZ}' + \mathbf{R} \tag{6.15}$$

where $\mathbf{G}$ is the variance matrix of $\mathbf{u}$ and $\mathbf{R}$ is the variance matrix of the residuals. $\mathbf{G} = \mathbf{A}\sigma_s^2$, where $\mathbf{A}$ = the numerator relationship matrix among the sires, and σ_s^2 = variance of the sire effect = ¼ of the additive genetic variance. The sire effect variance is equal to one-quarter of the additive genetic variance, because each sire passes half of his genes to each daughter. When squared to compute the variance, the one-half additive genetic effect becomes one-quarter of the additive genetic variance. Both $\mathbf{A}$ and $\mathbf{G}$ are always symmetrical matrices, that is, $\mathbf{G} = \mathbf{G}'$ and $\mathbf{A} = \mathbf{A}'$.

The diagonal elements of $\mathbf{A}$ will be equal to unity, because an individual has all of its genes in common with itself. The off-diagonal elements will reflect the fraction of genes that the two individuals, corresponding to the appropriate row and column of $\mathbf{A}$, have identical by descent, for example, 0.5 for father and son and 0.25 for half-sibs. The diagonal elements of $\mathbf{A}$ will be greater than unity for inbred individuals. This is because the genetic variance among a sample of inbred individuals will be greater than the variance among a sample of noninbred individuals.

As in the fixed model, the residuals will generally be assumed to be uncorrelated and have equal variance. In this case $\mathbf{R} = \mathbf{I}\sigma_e^2$, where $\mathbf{I}$ is the the identity matrix and σ_e^2 is the residual variance.

The Mixed Model Equations

We will now differentiate between the vector of fixed effects, $\boldsymbol{\beta}$, defined in Equation (6.14) and the solutions for the fixed effects which will be denoted as $\hat{\boldsymbol{\beta}}$. Similarly the solutions to the random effects will be denoted $\hat{\mathbf{u}}$. Henderson (1973) showed that solutions for the random effects can then be computed as $\mathbf{GZ'V}^{-1}(\mathbf{y}-\hat{\boldsymbol{\beta}})$. However, the variance matrix, $\mathbf{V}=\mathbf{ZGZ'}+\mathbf{R}$, is not diagonal and therefore cannot be inverted for very large data sets. Solutions for $\boldsymbol{\beta}$ and $\mathbf{u}$ for large data sets can be derived by solving the following set of equations, denoted the "mixed model equations" (Henderson, 1973):

$$\begin{bmatrix} \mathbf{X'R}^{-1}\mathbf{X} & \mathbf{X'R}^{-1}\mathbf{Z} \\ \mathbf{Z'R}^{-1}\mathbf{X} & \mathbf{Z'R}^{-1}\mathbf{Z}+\mathbf{G}^{-1} \end{bmatrix} \begin{bmatrix} \hat{\boldsymbol{\beta}} \\ \hat{\mathbf{u}} \end{bmatrix} = \begin{bmatrix} \mathbf{X'R}^{-1}\mathbf{y} \\ \mathbf{Z'R}^{-1}\mathbf{y} \end{bmatrix} \tag{6.16}$$

where $\mathbf{R}^{-1}$ is the inverse of the residual variance matrix and $\mathbf{G}^{-1}$ is the inverse of the variance matrix for $\mathbf{u}$. The left-hand side of these equations consists of a square symmetrical matrix termed the "coefficient matrix" and $\hat{\boldsymbol{\beta}}\hat{\mathbf{u}}$, the vector of solutions. As noted previously, for analysis of a single trait, it is generally assumed that the residual variances for each record are equal and uncorrelated. In this case, the residual variance matrix is equal to $\mathbf{I}\sigma_e^2$, and $\mathbf{R}^{-1}=\mathbf{I}/\sigma_e^2$. Thus the mixed model equations can be simplified by multiplying both sides by σ_e^2 as follows:

$$\begin{bmatrix} \mathbf{X'X} & \mathbf{X'Z} \\ \mathbf{Z'X} & \mathbf{Z'Z}+\mathbf{G}^{-1}\sigma_e^2 \end{bmatrix} \begin{bmatrix} \hat{\boldsymbol{\beta}} \\ \hat{\mathbf{u}} \end{bmatrix} = \begin{bmatrix} \mathbf{X'y} \\ \mathbf{Z'y} \end{bmatrix} \tag{6.17}$$

For the "sire model" given in Equation (6.13), $\mathbf{X'X}$ will be a diagonal matrix with rows and columns equal to the number of herds. The diagonal element of each row will be the number of records in the corresponding herd, and all off-diagonal elements will be zero. Similarly, $\mathbf{Z'Z}$ will be diagonal with each diagonal element equal to the number of daughter records of each sire. $\mathbf{X'Z}$ will have rows equal to the number of herds and columns equal to the number of sires. Each element will be the number of records in the corresponding herd × sire combination. $\mathbf{X'y}$ will be a vector of length equal to the number of herds, and each element will be the sum of the record values in the corresponding herd. The length of $\mathbf{Z'y}$ will be the number of sires, and each element will be the sum of the records of all the daughters of the corresponding sire.

Henderson (1973) termed the solutions of random effects in the mixed model "best linear unbiased predictors" (BLUP). Under the assumed variance structure, the random solutions in the mixed model equations, $\hat{\mathbf{u}}$, will be "best" in the sense that $E(\hat{\mathbf{u}}-\mathbf{u})^2$ will be minimized, within the assumed constraints. Since the random effects are not parameters, their solutions were termed "predictors" rather than "estimates."

Solving the Mixed Model Equations

Solutions to the mixed model equations can be obtained by multiplying the right-hand side vector by the inverse of the coefficient matrix. An exact solution requires inverting the coefficient matrix, but for simple single-trait models, such as in Equation (6.13), the coefficient matrix will be much smaller than $\mathbf{V}$.

If many effects are included in the model, approximate solutions can be obtained by iteration. There are several iteration methods that can be applied to solve the mixed model equations. Gauss–Seidel iteration is generally the method of choice, because it is relatively rapid and guaranteed to converge, provided the equations have a solution (Quaas and Pollak, 1980). The algorithm for Gauss–Seidel iteration is as follows: Define d_{ij} as an element ij of the coefficient matrix where i is the row index and j is the column index, x_i the solution for row i, and r_i the ith row of the right-hand side. Then the solution for x_i for the kth iteration is computed as follows:

$$x_i^k = \frac{r_i - \sum_{1}^{j-1}\left(d_{ij}x_j^{k-1}\right) + \sum_{j+1}^{J}\left(d_{ij}x_j^{k-1}\right)}{d_{ii}} \tag{6.18}$$

where J is the total number of columns. For the first round of iteration some initial guess for the values of $\mathbf{x}$ can be used, but generally $\mathbf{x}^1 = 0$ is assumed.

Gauss–Seidel iteration is guaranteed to converge to a solution, provided a solution exists. Convergence rate will depend on the ratio of diagonal to off-diagonal elements. Thus, sire models may converge in less than 10 rounds of iteration, because the diagonal elements are generally very large compared to the off-diagonal elements. Animal models, described later in the section: "The Individual Animal Model", generally required hundreds of rounds of iteration to achieve approximate convergence. The method of "preconditioned conjugate gradient" generally converges much faster for complicated models than Gauss–Seidel or other methods of iteration (Tsuruta *et al.*, 2001).

Computing the coefficient matrix still requires inverting $\mathbf{G}$. For a sire model $\mathbf{G}^{-1} = \mathbf{A}^{-1}/\sigma_s^2$, and $\mathbf{G}^{-1}\sigma_e^2 = \mathbf{A}^{-1}\sigma_e^2/\sigma_s^2$. σ_e^2/σ_s^2 is a constant, which is generally assumed known. Henderson (1976) developed a simple algorithm to invert $\mathbf{A}$ from a list of individuals and their sires and dams. Thus, the only matrix that must be inverted is the coefficient matrix, which will be a square matrix of size equal to the number of effects included in the model.

Important Properties of Mixed Model Solutions

The prediction error variances (PEV) of the fixed and random effects can be estimated by inverting the coefficient matrix of the mixed model equations. This inverse can be partitioned into four submatrices corresponding to the four submatrices in the mixed model equations. That is,

$$\begin{bmatrix} \mathbf{X'R^{-1}X} & \mathbf{X'R^{-1}Z} \\ \mathbf{Z'R^{-1}X} & \mathbf{Z'R^{-1}Z+G^{-1}} \end{bmatrix}^{-1} = \begin{bmatrix} \mathbf{C}_{11} & \mathbf{C}_{12} \\ \mathbf{C}_{21} & \mathbf{C}_{22} \end{bmatrix} \tag{6.19}$$

The diagonal elements of $\mathbf{C}_{11}$ will correspond to the PEV for the fixed effect solutions, and the diagonal elements of $\mathbf{C}_{22}$ will correspond to the PEV for the random effect solutions. Solutions for fixed effects will have greater variance than the actual effects, while PEV of the random effect solutions, which are regressed toward the mean, will be less than the variance of the effects. In general

$$\mathrm{Var}(\mathbf{u}) = \mathrm{Var}(\hat{\mathbf{u}}) + \mathrm{PEV}(\hat{\mathbf{u}}) \tag{6.20}$$

where Var($\mathbf{u}$) and Var($\hat{\mathbf{u}}$) are the variances of $\mathbf{u}$ and $\hat{\mathbf{u}}$ and PEV($\hat{\mathbf{u}}$) is the prediction error variance of $\hat{\mathbf{u}}$. Henderson (1973) also showed that the covariance of $\mathbf{u}$ and $\hat{\mathbf{u}}$ is equal to Var($\hat{\mathbf{u}}$). Thus, the regression of $\mathbf{u}$ on $\hat{\mathbf{u}}$ is equal to unity. That is, if the actual difference between two random effects is equal to x, the expected difference between their solutions will also be equal to x. This is not the case for fixed effects. The ratio Var($\hat{\mathbf{u}}$)/Var($\mathbf{u}$) is called the "reliability" of $\mathbf{u}$ and is equal to the square of the correlation between $\hat{\mathbf{u}}$ and $\mathbf{u}$, which is termed the coefficient of determination. The square root of the reliability is denoted the "accuracy" of $\mathbf{u}$.

Multivariate Mixed Model Analysis

The mixed model equations can also be used to analyze several correlated traits, for example, milk and butterfat production of cows. A multitrait sire model can be described as follows:

$$Y_{ijkl} = H_{il} + S_{jl} + e_{ijkl} \tag{6.21}$$

where Y_{ijkl} is the production record of cow k in herd i for trait l, H_{il} is the effect of herd i on trait l, S_{jl} is the effect of the cow's sire j on trait l, and e_{ijkl} is the random residual associated with trait l. In this case it will generally be assumed that both the additive genetic effects and the residuals have a multivariate normal distribution. As in the univariate case, the distribution of each record will be given by the distribution of the random genetic effect times the residual effect distribution. For two correlated traits, x and y, the distribution of the residuals for each individual will be as follows:

$$\left[2\pi\sigma_x^2\sigma_y^2\left(1-\rho^2\right)\right]^{-1/2} e^{\varphi} \tag{6.22}$$

where σ_x^2 and σ_y^2 are the residual variances for traits x and y, $\rho=\sigma_{xy}/\sigma_x\sigma_y$ is the residual correlation, and φ is computed as follows:

$$\varphi = \frac{1}{2\left(1-\rho^2\right)}\left[\frac{x-\mu_x^2}{\sigma_x^2} - 2\rho\frac{\left(x-\mu_x\right)^2\left(y-\mu_y\right)^2}{\sigma_x^2\sigma_y^2} + \frac{x-\mu_y^2}{\sigma_y^2}\right] \tag{6.23}$$

where μ_x and μ_y are the means for traits x and y and are equal to $H_{il}+S_{jl}$ for each trait. The distributions for the genetic effects are computed in a similar manner.

The residual variance matrix in the mixed model equations will no longer be diagonal, but will be "block diagonal." For two traits, the residual matrix will have the structure $\mathbf{I}\otimes\mathbf{R}_i$, where $\mathbf{I}$ is an identity matrix and $\mathbf{R}_i$ is a 2×2 matrix with elements as follows:

$$\mathbf{R}_i = \begin{bmatrix} \sigma_x^2 & \sigma_{xy} \\ \sigma_{xy} & \sigma_y^2 \end{bmatrix} \tag{6.24}$$

"$\otimes$" denotes the "Kronecker product," which means that each element of $\mathbf{I}$ is multiplied by $\mathbf{R}_i$. Similarly the variance matrix of the sire effect will be $\mathbf{A}\otimes\mathbf{S}$, where $\mathbf{S}$ is a 2×2 matrix as follows:

$$\mathbf{S} = \begin{bmatrix} \sigma_{sx}^2 & \sigma_{sxy} \\ \sigma_{sxy} & \sigma_{sy}^2 \end{bmatrix}$$

where σ_{sx}^2 and σ_{sy}^2 are the sire effect variances for traits x and y and σ_{sxy} is the covariance between them. Although both the residual and sire effect matrices can be easily inverted, the simplification obtained in Equation (6.17) on multiplying by the residual variance is no longer possible. The total number of equations will be the number of level of effects times the number of traits.

The Individual Animal Model

Henderson (1973) first proposed that the mixed model equations could be used to estimate polygenic breeding values for all animals in a population accounting for all known relationships, via the "individual animal model" (IAM). A simple IAM is given below:

$$Y_{ijk} = H_i + a_j + p_j + e_{ijk} \tag{6.25}$$

where Y_{ijk} is record k of individual j in "herd" or "block" i, H_i is the fixed effect of herd i, a_j is the random additive genetic effect of individual j, p_j is the random permanent environmental effect for individual j, and e_{ijk} is the random residual associated with each record. A permanent environmental effect is required if individuals can have multiple records, because there will generally be an effect common to all records of each individual in addition to the additive genetic effect common to all records of the individual. In matrix notation this model can be written as follows:

$$\mathbf{y} = \mathbf{X\beta} + \mathbf{Z}_1\mathbf{a} + \mathbf{Z}_2\mathbf{p} + \mathbf{e} \tag{6.26}$$

where $\mathbf{Z}_1$ is the coefficient matrix for the additive genetic effects, $\mathbf{Z}_2$ is the coefficient matrix for the permanent environmental effects, and $\mathbf{a}$ and $\mathbf{p}$ are the vectors of additive genetic and permanent environmental effects, respectively.

In a fixed model, the additive genetic and permanent environmental effects would be completely confounded, because each level of these two effects refers to the same individual. In the IAM these effects can be estimated separately, because both are assumed to be random and their variance structures are different. The variance matrix for the permanent environmental effect will be $\mathbf{I}\sigma_p^2$, where σ_p^2 is the variance component of the permanent environmental effect. The variance matrix for the additive genetic effect will be $\mathbf{A}\sigma_a^2$, where σ_a^2 is the additive genetic variance. After multiplying by the residual variance, the mixed model equations for this model are

$$\begin{bmatrix} \mathbf{X'X} & \mathbf{X'Z}_1 & \mathbf{X'Z}_2 \\ \mathbf{Z}_1'\mathbf{X} & \mathbf{Z}_1'\mathbf{Z}_1 + \mathbf{G}^{-1}\sigma_e^2 & \mathbf{Z}_1'\mathbf{Z}_2 \\ \mathbf{Z}_2'\mathbf{X} & \mathbf{Z}_2'\mathbf{Z}_1 & \mathbf{Z}_2'\mathbf{Z}_2 + \mathbf{I}\gamma \end{bmatrix} \begin{bmatrix} \hat{\boldsymbol{\beta}} \\ \hat{\mathbf{a}} \\ \hat{\mathbf{p}} \end{bmatrix} = \begin{bmatrix} \mathbf{X'y} \\ \mathbf{Z}_1'\mathbf{y} \\ \mathbf{Z}_2\mathbf{y} \end{bmatrix} \tag{6.27}$$

where $\gamma = \sigma_e^2/\sigma_p^2$.

In most cases not all individuals included in the population will have records. For example, if the trait analyzed is milk production, only females will have production records. Individuals without records, such as sires of cows, can be included in the analysis via the relationship matrix. Additional equations can be added for these individuals in the mixed model equations. For these animals all elements of the mixed model equations will be zero, except for those included in $\mathbf{G}^{-1}$. Thus in the IAM sire evaluations are derived via the relationship matrix.

The animal model as described would accurately reflect reality if the pedigree of all animals could be traced back to a group of unrelated animals with assumed equal genetic value. However, this is never the case. Various animals of different ages will be missing pedigree information. Thus a fixed "genetic group" effect is usually included in the model to account for genetic trend not included in the known genetic relationships. Thompson (1979) and Robinson (1986) proposed a grouping strategy based on "phantom parents." Each individual with unknown parents is assigned phantom parents. These phantom parents are then assigned into groups based on year of birth, sex, and whether the sire, dam, or both parents are unknown. The genetic evaluations are then computed as the sum of the additive genetic effects and the group effects of each individual. Westell *et al.* (1988) developed a simple algorithm to directly compute estimated breeding values that incorporate the group effects for each individual.

Even though this system of equations will generally be very large, it will also be quite "sparse," that is, more than 90% of the total number of elements in the coefficient matrix will be equal to zero. These equations can also be solved by Gauss–Seidel iteration. As noted earlier, the number of iterations required for convergence is a function of the size of the diagonal elements in the coefficient matrix compared to the off-diagonal elements. In sire models diagonal elements are generally quite large, because each sire has many daughters with records. This is not the case for animal models. In the IAM, the diagonal element consists only of the contribution of the inverse of the relationship matrix, plus the individual's own records. Thus, many more iterations will be required in the IAM, as compared to sire models in which the diagonal elements are generally much greater than the off-diagonal elements.

Theoretically it will still be possible to obtain PEV of the genetic evaluations of all animals via inversion of the coefficient matrix, but this will generally not be possible in practice. Relatively simple algorithms have been developed to derive approximate PEV which were shown by simulation to be very close to the true values (Misztal and Wiggans, 1988).

Yield Deviations and Daughter Yield Deviations

Animal model evaluations combine information from an animal and all relatives, but the algebra required to derive solutions can be explained easily without recourse to matrix algebra (VanRaden and Wiggans, 1991). Consider the following single-trait animal model given in matrix notation:

$$\mathbf{y} = \mathbf{Mm} + \mathbf{Za} + \mathbf{ZA_g g} + \mathbf{Pp} + \mathbf{e}$$

where $\mathbf{y}$ represents the production record; $\mathbf{m}$, $\mathbf{a}$, $\mathbf{g}$, and $\mathbf{p}$ are the vectors of effects for management group, random portion of additive genetic merit, unknown-parent group, and permanent environment, respectively; $\mathbf{M}$, $\mathbf{Z}$, $\mathbf{ZA_g}$, and $\mathbf{P}$ are incidence matrices for these effects; and $\mathbf{e}$ is the residual variance. The matrix $\mathbf{A_g}$ relates animals to unknown-ancestor groups, as described in the previous section.

The cow's own information is summarized by her "yield deviation" (YD), a weighted average of yields adjusted for effects other than genetic merit and error. Defining $\hat{\mathbf{m}}$ and $\hat{\mathbf{p}}$ as the vectors of solutions for $\mathbf{m}$ and $\mathbf{p}$, each cow's YD is computed as the element of $\mathbf{Z}'(\mathbf{y}-\mathbf{M}\hat{\mathbf{m}}=\mathbf{P}\hat{\mathbf{p}})$ for that cow divided by the corresponding diagonal element of $\mathbf{Z}'\mathbf{Z}$; that is, a weighted average of the cow's yields adjusted for effects other than genetic merit and residual.

For each bull with daughter records in the analysis, the bull's daughter yield deviation (DYD) is computed by summing over all daughters with records as follows:

$$\text{DYD} = \frac{\sum_1^N \left[q_{\text{prog}} \text{w2}_{\text{prog}} \left(\text{YD}_{\text{prog}} - \text{PTA}_{\text{mate}} \right) \right]}{\sum_1^N q_{\text{prog}} \text{w2}_{\text{prog}}} \tag{6.28}$$

where N is the number of daughters of each sire. For each daughter with records, q_{prog} equals 1 if progeny's other parent is known and 2/3 if unknown, where $w2_{prog}$ is the the number of lactations divided by the sum of the number of lactations and $2q_{prog}(\sigma_e^2/\sigma_a^2)$, YD_{prog} is the daughter's YD, and PTA_{mate} is the predicted transmitting ability of the cow's dam, the bull's mate. PTA is half of the animal's breeding value. Thus a DYD is the weighted mean of the bull's daughters' YD corrected for the genetic merit of the daughters' dams. (This definition of DYD is somewhat simplified, as compared to the definition of VanRaden and Wiggans (1991), based on the assumption that all lactation records are weighted equally.)

Analysis of DYD as the Dependent Variable

At present, and in the foreseeable future, only a small fraction of a commercial population will be genotyped for genetic markers. Analysis of the records of only those individuals that were genotyped will be problematic, because it will generally not be possible to estimate "nuisance" effects, such as herd or parity from the truncated sample of genotyped animals.

VanRaden and Wiggans (1991) wrote: "The DYD may be helpful in explaining evaluations and also as a dependent variable in statistical tests and calculation of conversions across countries." Unlike a simple mean of the cows' records or a simple mean of bull's daughter records, YD and DYD are corrected for the other effects included in the model. However, unlike genetic evaluations, both YD and DYD are not regressed toward the mean, as a function of the trait heritability.

Analysis of DYD as the dependent variable has the advantage, compared to genetic evaluations, that the variance of DYD decreases with the number of actual records included in the bull's DYD. Thus weighting DYD by the reliabilities of the genetic evaluations should yield "reasonable" results. That is, more weight would be given to DYD with lower variances. (The weighting factor of a record should be inversely related to the residual variance.)

Georges *et al.* (1995) first proposed that DYD should be used as the dependent variable in the analysis of the effects of genetic markers on quantitative traits in dairy cattle when only sires are genotyped, and the vast majority of published studies based on analysis of bull genotypes have used either DYD or "deregressed" genetic evaluations as the dependent variable. (Equations to computed deregressed genetic evaluations will be presented in Chapter 16.) It was assumed that effects estimated by analysis of genetic evaluations would be biased, due to the fact that the genetic evaluations are regressed toward the mean, while this would not be the case for DYD. This assumption was tested by Israel and Weller (1998). They found that estimates of candidate gene effects as derived from analysis of either DYD or sire genetic evaluations were underestimated, but less so by analysis of DYD.

A problem with the analysis of DYD as the dependent variable is the appropriate residual matrix structure. As noted previously for single-trait analyses, it is generally assumed that the residual matrix is equal to the identity matrix times a constant. This will not be the case for DYD, since DYD of related bulls will have a positive covariance. Various studies have therefore assumed that a variance matrix equal to the relationship matrix among bulls times a constant. In addition the residuals will be a function of the number of daughter records per bull. The standard procedure to account for this is to weight the DYD by the reliabilities of the evaluations. Although this may be approximately correct, over the general range of sire reliabilities, it is not clear what should be done with bulls with thousands of daughters. These bulls should have residual variances approaching zero, and it is not clear how the generalized linear model should behave in this situation.

Summary

Analysis of mixed models is much more complicated than analysis of fixed models. These models are preferred for genetic analysis, first because they utilize information on genetic variance and genetic relationships among animals that cannot be utilized by fixed models. Unlike estimation of fixed effects, random effect solutions are regressed toward the mean. That is, the variances of the solutions increase as a function of the quantity of information included in the analysis. This allows for accurate comparison of the genetic values of individuals with widely differing amounts of information. For fixed model solutions the variances decrease as the amount of information increases. The most common models for genetic evaluation in dairy cattle are single- and multitrait animal models. The cow's own information is summarized by her YD, a weighted average of yields adjusted for effects other than genetic merit and error. DYD are the weighted mean of the bulls' daughters' YD corrected for the genetic merit of the daughters' dams. The vast majority of published studies based on analysis of bull genotypes have used either DYD or "deregressed" genetic evaluations as the dependent variable, and these models will be considered in detail beginning in Chapter 16.

7 Maximum Likelihood and Bayesian Estimation of QTL Parameters with Random Effects Included in the Model

Introduction

Only fixed effects should be estimated by maximum likelihood estimation (MLE), while random effects should be "removed" by integration (Titterington *et al.*, 1985). In the estimation of QTL effects from segregating populations, it will generally be necessary to include polygenic random effects in the model. In this chapter we will present likelihood equations for two QTL estimation designs that require inclusion of random polygenic effects in the analysis model: the "daughter" and "granddaughter" designs. Although the likelihood equations cannot be solved analytically, they can be solved by iterative methods. In the last sections we will consider Bayesian estimation of the parameters of the granddaughter design, based on assumptions with respect to the prior distributions of the QTL parameters.

Maximum Likelihood Estimation of QTL Effects with Random Effects Included in the Model, the Daughter Design

Estimation of QTL effects with random polygenic effects included in the model will be illustrated using the "daughter design," first proposed by Neimann-Sørensen and Robertson (1961). The daughter design for a single family is illustrated in Figure 7.1. The daughters of a sire known to be heterozygous for a genetic marker are genotyped for the marker and scored for the quantitative trait. Since the dam genotypes are generally unknown and differ among individuals, the dam alleles for the marker locus and QTL are denoted M_x and Q_x, respectively.

If we assume that only the same two QTL alleles are present in the dam population, with frequencies of p and $1-p$, then the contrast between the two groups of progeny will be $a(1-2r)+d(1-2r)(1-2p)$, where a and d are the additive and dominance effects and r is the recombination frequency between the two loci. These parameters are confounded in a linear model analysis.

Even if a QTL is segregating in the population, a specific parent may be homozygous for the QTL. Therefore, most studies have been based on analysis of several heterozygous parents. Even if some of the individuals analyzed are heterozygous for a marker-linked QTL, the linkage relationships may be different for different individuals. Thus, summed over all progeny groups,

Genomic Selection in Animals, First Edition. Joel Ira Weller.

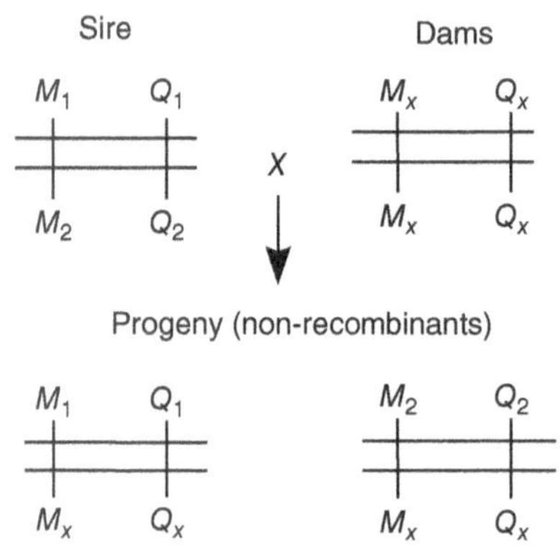

Figure 7.1 The daughter design.

there may be no effect associated with the segregating marker alleles. The appropriate linear model for the daughter design with multiple families is therefore

$$Y_{ijkl} = S_i + M_{ij} + B_k + e_{ijkl} \tag{7.1}$$

where S_i is the effect of the ith parent, M_{ij} is the effect of the jth allele, nested within the ith parent, B_k is the effect of the kth herd, and e_{ijkl} is the random residual.

The progeny group inheriting sire allele M_1 and the group inheriting sire allele M_2 are compared. If the assumptions listed earlier hold, and if the distribution of dams between the two groups is random, then a difference between the two groups of progeny for the quantitative trait will be due to a QTL linked to M heterozygous in the sire. This assumes that marker allele origin can be determined for the daughters. Significance of a segregating QTL linked to the genetic marker can be tested by ANOVA. Under the null hypothesis of no segregating QTL, the ratio of the marker allele effect mean squares to the residual mean squares should have a central F-distribution.

The likelihood function for the daughter design including a polygenic sire effect is (Weller, 2009)

$$L = \prod_{k=1}^{K} \int \left\{ f\left(g_k - \mu_g, \sigma_g^2\right) \sum_{v=1}^{4} P_v \prod_{i=1}^{3} \prod_{l-1}^{L_i} \sum_{j-1}^{3} c_{j|i,v} f\left(y_{ikl} - \mu_j - g_k, \sigma_e^2\right) \right\} dg_k \tag{7.2}$$

where K is the number of sires, P_v is the probability of sire QTL genotype v, $c_{j|i,v}$ is the probability of progeny QTL genotype j conditional on the combination of sire QTL genotype v and progeny marker genotype i, L_i is the number of daughters with marker genotype i, y_{ikl} is the trait value for progeny l of sire k, with marker genotype i, and $f(g_k - \mu_g, \sigma_g^2)$ represents the normal density function for the sire effects. This function has a mean of μ_g and a variance of σ_g^2, which will be equal to one-quarter of the additive genetic variance not explained by the segregating QTL. $f(y_{ikl} - \mu_j - g_k, \sigma_e^2)$ represents a normal density function with a mean of $\mu_j + g_k$ and a variance of σ_e^2. σ_e^2 includes the residual variance, and three-quarters of the genetic variance not explained by the segregating QTL. The likelihood function is the joint density of the observations, integrated over g_k, the polygenic sire effect, which is assumed to be random.

Although the integral cannot be solved analytically, it can be approximately solved by summation of each sire. Thus, the likelihood value can be approximately computed for any combination of parameter values. However, this model still does not include "nuisance" fixed effects, such as herd-year-season

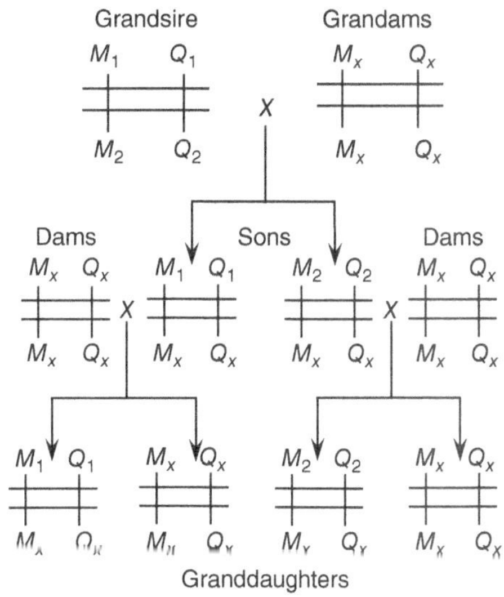

Figure 7.2 The granddaughter design.

effects, which would have to be included in any analysis of field data. Therefore, it is not surprising that ML solutions have not been computed on actual data. Although instead of analyzing actual production records, it would be possible to analyze "yield deviations," which are means of cow records corrected for fixed effects (VanRaden and Wiggans, 1991). For the granddaughter design it is necessary to include normal density distribution terms for both the sire and grandsire, and integrate over both terms, if the analysis is based on the actual production records.

The Granddaughter Design

Hoeschele and VanRaden (1993a, 1993b) derived Bayesian estimates of QTL parameters for the "granddaughter design" (Weller *et al.*, 1990). Most studies that have attempted to estimate QTL effects in dairy cattle via linkage to microsatellites have used this design, diagrammed in Figure 7.2. Sons of grandsires heterozygous for the genetic markers are genotyped, and the daughters of these sons (i.e., the granddaughters of the original grandsire) are scored for the quantitative traits. It is assumed that both grandsire and son mates are random. The expectation of the contrast between the grandprogeny groups is only half as large as for the daughter design. However, the much greater number of phenotypic records can more than compensate for the reduction in the contrast.

This design has the advantage that for certain species, especially dairy cattle, the commercial population has the appropriate population structure, and records on quantitative traits of interest are recorded by the industry. Furthermore, it may be logistically easier to obtain biological material from AI sires which are located at a few AI centers, rather than cows that are scattered over a large number of herds. A segregating marker-linked QTL can be detected with analysis by the following linear model:

$$Y_{ijklm} = GS_i + M_{ij} + SO_{ijk} + B_l + e_{ijklm} \tag{7.3}$$

where GS_i is the effect of the ith grandparent, SO_{ijk} is the effect of the kth son with the jth marker allele, progeny of the ith grandparent, and the other terms are as defined for Equation (7.1). As in the daughter design, a significant marker-allele effect will be indicative of a linked QTL. Significance of this effect can be tested by ANOVA, with the marker mean squares in the numerator. However, this mean square will also include a component due to differences among sons. Thus, the denominator for the appropriate *F*-statistic will be a function of both the sons and residual mean squares (Ron *et al.*, 1994).

It is not possible with the granddaughter designs to estimate dominance at the QTL. Even if the actual QTL alleles are identified, and both grandsires and granddams are genotyped, the expectation of the effect of the sons that are heterozygous for the QTL will still be midway between the two homozygous groups.

As noted earlier for the daughter design, this model is generally not applied, because daughters have multiple records and it will not be possible to accurately estimate fixed effects. Generally either the genetic evaluations or the daughter yield deviations (DYD) of the sons (VanRaden and Wiggans, 1991) will be analyzed. In this case there is only a single record for each son, and the analysis model is as follows:

$$Y_{ijk} = GS_i + M_{ij} + e_{ijk} \tag{7.4}$$

where Y_{ijk} is the genetic evaluations or DYD for son k of grandsire i that received grandpaternal allele j, and the other terms are as defined in the previous equation.

Determination of Prior Distributions of the QTL Parameters for the Granddaughter Design

In order to derive a prior distribution of QTL parameters, it is necessary to make assumptions about the relevant QTL parameters: the QTL genotype means and variances, the number of frequencies of QTL alleles, and the QTL location. Hoeschele and VanRaden (1993a) simplified the analysis somewhat by employing the following assumptions:

1. For each QTL only two alleles are segregating in the population.
2. All QTL were assumed codominant. Strictly speaking this assumption is not required for a granddaughter design, because only substitution effects are estimable.
3. The residual variance is independent of the QTL genotypes.

Under these assumptions, prior distributions must be derived for only three parameters: the QTL additive effect, the allele frequency, and the QTL location. No prior assumptions are required with respect to the residual variance, which is also estimated, and the total additive genetic variance including the segregating QTL, σ_A^2, is assumed to be known without error.

Although the actual distribution of QTL effects is unknown, it is known that the total variance contributed by all QTL should be no larger than σ_A^2. Most simulation studies have assumed that polygenic variance is due to a few QTL with relative large effects and numerous QTL with progressively smaller effects. Several mathematical models that generate this type of distribution have been proposed, and these models will be considered in Chapter 9. Hoeschele and VanRaden (1993a) assumed a prior exponential distribution of QTL effects. The exponential distribution has the form

$$f(a) = \lambda e^{-\lambda a} \tag{7.5}$$

where a is the QTL additive effect and λ is the parameter of this distribution. The statistical density of this distribution is maximum with $a=0$ and is equal to λ. The expectation of the distribution, that is, the expectation of a, is $1/\lambda$.

Although the additive effect can have a value from zero to infinity, Hoeschele and VanRaden (1993a) imposed lower and upper bounds. A lower bound was imposed, because very small QTL cannot be detected by the sample sizes generally considered. An upper bound was imposed for two reasons. First, a very large additive effect will lie outside the permissible parameter space, determined by σ_A^2. In this case the QTL will explain more than the total genetic variance, unless the allelic frequency is very low. Second, with values of λ that are appropriate for polygenic inheritance, the probability of sampling a very large effect tends toward zero and can therefore be ignored. Therefore, this density function must be divided by a constant to account for the extremes of the theoretical exponential distribution that are deleted from consideration. The value of this constant is $e^{-\lambda a l} - e^{-\lambda a u}$, where a_l and a_u are the lower and upper limits of a, respectively.

As noted earlier, Hoeschele and VanRaden (1993a) assumed only two alleles for each QTL segregating in the population. Thus, it is necessary to determine a prior distribution of the allelic frequency for only one allele. Hoeschele and VanRaden (1993a) assumed a uniform distribution over the range of zero to unity, subject to two restrictions. First, the frequency of the less frequent allele must be high enough, so that at least one of the sires included in the analysis is heterozygous for the QTL. This will be considered again later. Second, the variance contributed by each QTL must be no greater than σ_A^2. Therefore, the joint distribution of the additive QTL effect, and allelic frequency, p, is

$$f(a,p) = \begin{cases} k^* f(a) & \text{if } 2p(1-p)a^2 \le \sigma_A^2 \\ 0 & \text{otherwise} \end{cases} \tag{7.6}$$

where k is the reciprocal of the integral of $f(a, p)$ over the restricted space of a and p.

The prior distribution for the QTL location parameter was computed based on the assumption of a uniform distribution throughout the genome. Two situations must be considered: linkage between the QTL and the genetic markers and nonlinkage. In the case of a single marker, nonlinkage can be defined as $r=0.5$, where r is the recombination frequency between the two loci. The joint prior density of a, p, and r can be represented as follows:

$$\text{Prior}(a,p,r) = \begin{cases} \text{Prob}(r=0.5) \\ \left[1-\text{Prob}(r=0.5)\right] * f(a,p) * f(r) \end{cases} \tag{7.7}$$

where $f(r)$ is the density of the distribution of r if the marker and the QTL are linked. If r is measured in Morgans, then $f(r)$ would have a uniform distribution. However, r is measured in recombination frequency, and r is therefore a nonlinear function of genetic map length for the commonly used mapping functions, such as Haldane or Kosambi. If $g(r)$ is the assumed mapping function, so that $\delta = g(r)$, where δ is the map distance between the QTL and the genetic marker, then

$$f(r) = \frac{f\left[g(r)\dfrac{d\delta}{dr}\right]}{\text{Prob}(\delta < \delta_r)} \tag{7.8}$$

where δ_r is the maximum linkage distance at which linkage can be detected in the same map units as δ.

If the genome consisted of a single circular chromosome, then the probability of linkage would be $(2\delta_r N_Q)/L_t$, where N_Q is the detectable number of segregating QTL and L_t is the total genome length, with both δ_r and L_t measured in genetic map units. For example, if $\delta_r = 1$ Morgan, and $N_Q = 10$, and $L_t = 30$ Morgans, then Prob($\mathbf{r}$=0.5) = 1 − 10/30 = 0.67. The detectable number of QTL, N_Q, was derived as follows:

$$N_Q = F^* \frac{\sigma_A^{\ 2}}{E(V_Q)} \tag{7.9}$$

where F is the fraction of the genome under analysis for QTL and $E(V_Q)$ is the expected variance due to a single detectable QTL, which is computed as follows:

$$E(V_Q) = k\int_{al}^{au} f(a)\left\{\int_0^{p_a}\left[2p(1-p)a^2\right]dp + \int_{1-p_a}^{1}\left[2p(1-p)a^2\right]dp\right\}da \tag{7.10}$$

where p_a is the appropriate value of p for each value of a.

As noted earlier, in the granddaughter design, p will be estimated as the frequency of one of the QTL alleles within the sample of grandsires. The total number of alleles will be twice the number of grandsires, and at least one of the grandsires must be heterozygous for the QTL. Therefore, the lower and upper bounds for p are $1/2G$ and $1-1/2G$, where G is the number of grandsires.

Setting a_l and a_u at approximately 0.2 and 1.1 genetic standard deviations, and $1/\lambda$ at 0.36 genetic standard deviations, $N_Q = 10$ for a complete genome scan of 10 grandsire families (Hoeschele and VanRaden, 1993a). With a heritability of 0.25, these limits for the additive effect are equal to 0.1 and 0.55 phenotypic standard deviations. More than 200 sons per sire will be required to obtain power greater than 0.5 to detect a QTL of 0.1 phenotypic standard deviations (Weller *et al.*, 1990).

With a single marker, and a genome divided into chromosomes of differing lengths, Prob(r=0.5) will also depend on the length of the marked chromosome and the position of the marker along the chromosome. If the marker is located at one end of the chromosome, then the length of the chromosomal segment for which a QTL can be detected is only δ_r, instead of $2\delta_r$. The final calculations for Prob(r=0.5) and $f(r)$, considering any possible marker location and genome and any mapping function, are rather complicated and are given in Hoeschele and VanRaden (1993a).

With a marker bracket, Prior(a, p, r) becomes

$$\text{Prior}(a,p,r) = \begin{cases} 1-\text{Prob}(0<r_1<0) \\ \left[\text{Prob}(0<r_1<R)\right]^* f(a,p)^* f(r_r) \end{cases} \tag{7.11}$$

where r_1 is the recombination frequency between one of the markers and the QTL, R is the recombination frequency between the two markers of the marker bracket, and

$$\text{Prob}(0<r_1<R) = \frac{\delta_R N_Q}{L_t} \tag{7.12}$$

with δ_R equal to the length of the marker bracket in map units, again assuming a uniform distribution for the QTL location. $f(r_1)$ is computed as $f(r)$ for a single marker.

Formula for Bayesian Estimation and Tests of Significance of a Segregating QTL in a Granddaughter Design

In order to derive Bayesian estimates for the QTL parameters, the prior density function is multiplied by the likelihood function. The likelihood function for the daughter design is given in Equation (7.2). This will be nearly the same function for the granddaughter design if the analysis is performed on DYD with a single record for each son. The only difference is that the residual variances of the DYD are not equal, but are a function of the number of daughters per son.

The posterior distribution of the QTL parameters given a single marker also consists of a discrete part, if the marker is not linked to a segregating QTL, and a continuous part, if a linked QTL is detected. The complete posterior distribution can be described as follows (Hoeschele and VanRaden, 1993a):

$$\text{Posterior}(a,p,r)=\begin{cases}\text{Prob}(r=0.5\mid \mathbf{y},\mathbf{M})\\ 1-\left[\text{Prob}(r=0.5\mid \mathbf{y},\mathbf{M})\right]*f(a,p,r\mid \mathbf{y},\mathbf{M})\end{cases} \tag{7.13}$$

where **y** and **M** represent the phenotypic and marker data, respectively. The posterior probability of no linkage is calculated as follows:

$$\text{Prob}(r=0.5|\mathbf{y},\mathbf{M})=\frac{\text{Prob}(r=0.5)E\left[L(r=0.5)\right]}{\text{Prob}(r=0.5)E\left[L(r=0.5)\right]+\left[1-\text{Prob}(r=0.5)\right]E\left[L(r<0.5)\right]} \tag{7.14}$$

where $E[L(r=0.5)]$ and $E[L(r<0.5)]$ are the expectations of the likelihood function with $r=0.5$ and $r<0.5$, respectively. $E[L(r<0.5)]$ is computed as follows:

$$E\left[L(r<0.5)\right]=\int_0^{0.5}\left[\int_{a_l}^{a_u}\left[\int_{p_l}^{p_u}L(\mathbf{y}|\mathbf{M};r,p,a)f(p,a)f(r)dp\right]da\right]dr \tag{7.15}$$

where $L(\mathbf{y}|\mathbf{M};\, r,\, p,\, a)$ is the likelihood function as computed previously. Similarly, $E[L(r=0.5)]$ is computed with r fixed at 0.5—that is, the expectation of the likelihood without a segregating QTL linked to the marker, which is a standard polygenic sire model. The posterior density of the QTL parameters is computed as follows:

$$f(r,p,a\mid \mathbf{y},\mathbf{M})=L(\mathbf{y}|\mathbf{M};r,p,a)\frac{f(a,p)f(r)}{f(\mathbf{y}|\mathbf{M})} \tag{7.16}$$

where $f(\mathbf{y}|\mathbf{M})$ is the denominator of Equation (7.14). Assuming a uniform loss function, the point estimates for r, p, and a are derived by maximizing the statistical density, that is, the mode of the distribution. With a quadratic loss function, the parameter estimates are derived by maximizing the mean of the distribution.

Linkage of the genetic marker to a segregating QTL can be tested by comparing the posterior probabilities of $r=0.5$ and the posterior probability that $r<0.5$. If both errors are of equal economic value, then the hypothesis of $r=0.5$ will be rejected if the posterior probability is less than half.

Summary

Equations were presented to derive estimates of QTL parameters for daughter and granddaughter designs by maximum likelihood and Bayesian methodology. In both cases a single segregating QTL linked to a single genetic marker was assumed. Thus the specific methods presented cannot be directly applied to genomic selection in which it will be necessary to derive estimates for thousands of potential QTL estimated from a dense marker map. The main usefulness of the equations presented in this chapter is therefore first historical and second to give an example of the types of solutions that are possible using these methodologies. Thus these methods can serve as an introduction to the methods presented in Chapter 16 for genomic analysis based on dense marker maps covering the entire genome.

8 Maximum Likelihood, Restricted Maximum Likelihood, and Bayesian Estimation for Mixed Models

Introduction

In the first section of this chapter we will demonstrate how solutions to the mixed model equations can be derived by maximum likelihood (ML). However, there are no simple algorithms for deriving ML parameter solutions for large mixed model systems. In practice solutions for fixed and random effects will be derived from the mixed model equations presented in Chapter 6, under the assumption that the variance components (VC) are known without error. In the following sections we will consider methodologies for estimation of VC, including analysis of variance and ML methods. Only Henderson's Method III yields unbiased estimates of VC. We will explain why estimates of VC derived by simple ML are biased and show how this bias can be corrected by application of "restricted maximum likelihood" (REML). (We should note that REML variance component estimates are still biased, but less so than simple ML estimates.) In the final section we will show how VC for the mixed model can be derived by Gibbs sampling (GS).

Derivation of Solutions to the Mixed Model Equations by Maximum Likelihood

For mixed models, the likelihood function is the joint density function integrated over the random effects (Titterington *et al.*, 1985). We will illustrate this based on the simple mixed model for genetic analysis of dairy cattle including a fixed herd effect and a random sire effect. The statistical density function for this model assuming unrelated sires will be

$$f\left(Y_{ijk}\right)=\prod^{J}\left\{f\left(s_j\right)\prod^{IK}f\left(Y_{ijk}\right)\right\} \tag{8.1}$$

where $f(Y_{ijk})$ is the density function for individual k, from herd i, daughter of sire j; $f(s_j)$ represents the density function for sire j, which is the normal density function with a mean of zero and a variance of σ_s^2; and $f(y_{ijk})$ represents the normal density function for daughter k of sire j, which has a mean of $s_j + h_i$ and a variance of σ_e^2. The likelihood function is then computed by integrating with respect to the random sire effects as follows:

$$L=\prod^{J}\int\left\{f\left(s_j\right)\prod^{IK}f\left(Y_{ijk}\right)\right\}ds_j \tag{8.2}$$

Genomic Selection in Animals, First Edition. Joel Ira Weller.

The ML parameter solutions are those values of the fixed effects and variances that maximize the likelihood function.

Estimation of the Mixed Model Variance Components

In order to solve the mixed model equations, the variances of the random effects must be known. In practice these variance components (VC) must be estimated from the same data. Variance component estimation will be considered here in only very general terms. Various methods have been proposed to estimate VC. These methods can be grouped into "analysis of variance" type methods, "maximum likelihood" type methods, and Bayesian type methods, specifically Gibbs sampling (GS), first considered in Chapter 5. In analysis of variance type methods, VC are estimated by first computing solutions for the fixed and random effects. The VC are then estimated by their expectations, which are functions of the solutions. Until the 1980s, Henderson's Method III was the most common method for variance component estimation. Although it was unbiased, it did not guarantee that the estimates obtained would be within the possible parameter space—that is, variances greater or equal to zero and correlations among VC within the range of −1 and 1. (There are additional conditions that must be met for a valid variance–covariance matrix but are beyond the scope of this text. See Henderson (1984).)

Unlike analysis of variance type methods, maximum likelihood (ML) and Bayesian methods require that the estimates be within the parameter space, since an estimate outside the parameter space has by definition a likelihood of zero. However, this property also means that the estimates cannot be completely unbiased. This can be explained as follows. Assume that the variance component for a random effect tends toward zero. Since the ML estimate cannot be negative, because a negative variance component is impossible, only positive estimates will be obtained, and the expectation of the estimates will be greater than 0.

Maximum Likelihood Estimation of Variance Components

We will first describe maximum likelihood estimation (MLE) of VC and then describe the modifications required for restricted ML estimation (REML). Both methods are by necessity iterative, because the formulas used to estimate the VC are functions of the mixed model solutions, which are computed based on the previous estimates of the parameter values. The derivation given here closely follows Lynch and Walsh (1998). For a more detailed explanation, see Searle *et al.* (1992).

As in the previous chapter, ML estimates are derived by constructing the likelihood function (the joint density function of the observations), differentiating the log of this function with respect to the parameters, setting these differentials equal to zero, and solving for the parameter values in the resultant system of equations. For the mixed model, the parameters are the VC and the fixed effects. The likelihood function is the joint density of all observations after integrating over the random effects.

As assumed previously, the distribution of y in the mixed model is multivariate normal. Accounting for the fact that the means and variances in the mixed model can be different for each observation, this likelihood becomes

$$L = \prod^{N}\left[\frac{e^{-(y_i-\mu_i)^2/2\sigma_i^2}}{\sqrt{2\pi\sigma_i^2}}\right] \tag{8.3}$$

where N is the sample size and μ_i and σ_i^2 are the mean and variance for observation i.

For any series of values, x_1 to x_n,

$$\prod[e^{x_i}] = e^{\Sigma x_i} \tag{8.4}$$

Therefore, after removing constants from the multiplicative sum, this equation becomes

$$L = 2\pi^{-N/2}\left(\prod\sigma_i^2\right)^{-1/2}\left[e^{\frac{-\Sigma(-y_i-\mu_i)^2}{2\sigma_i^2}}\right] \tag{8.5}$$

Since $\prod\sigma_i^2 = |\mathbf{V}|$, where $|\mathbf{V}|$ is the determinant of $\mathbf{V}$ (Searle, 1982), this equation can be written in matrix notation as follows:

$$L = 2\pi^{-N/2}|\mathbf{V}|^{-1/2}\left[e^{-1/2(\mathbf{y}-\mathbf{X}\beta)'\mathbf{V}^{-1}(\mathbf{y}-\mathbf{X}\beta)}\right] \tag{8.6}$$

The natural log of the likelihood function is as follows:

$$\text{Log } L = -\left(\frac{N}{2}\right)\ln(2\pi) - \frac{1}{2}\ln|\mathbf{V}| - \frac{1}{2}(\mathbf{y}-\mathbf{X}\boldsymbol{\beta})'\mathbf{V}^{-1}(\mathbf{y}-\mathbf{X}\boldsymbol{\beta}) \tag{8.7}$$

For a mixed model in which the random effects are additive genetic effects, $\mathbf{V}$ is computed as follows:

$$\mathbf{V} = \mathbf{ZGZ}' + \mathbf{R} = \mathbf{Z}\left(\mathbf{A}\sigma_u^2\right)\mathbf{Z}' + \mathbf{I}\sigma_e^2 \tag{8.8}$$

In theory, ML estimates can now be derived by differentiating the right-hand side of Log L with respect to $\boldsymbol{\beta}$, σ_u^2, and σ_e^2 and setting these derivatives equal to zero. However, the ML solutions for $\boldsymbol{\beta}$ are themselves functions of the VC. Therefore an iterative solution will be necessary. Differentiating Log L with respect to $\boldsymbol{\beta}$ gives

$$\frac{\partial(\text{Log } L)}{\partial\boldsymbol{\beta}} = -2\mathbf{X}'\mathbf{V}^{-1}(y - \mathbf{X}\boldsymbol{\beta}) \tag{8.9}$$

Setting this derivative equal to zero and solving for $\boldsymbol{\beta}$ give

$$\hat{\boldsymbol{\beta}} = \left(\mathbf{X}'\hat{\mathbf{V}}^{-1}\mathbf{X}\right)^{-1}\mathbf{X}'\hat{\mathbf{V}}^{-1}\mathbf{y} \tag{8.10}$$

where $\hat{\boldsymbol{\beta}}$ and $\hat{\mathbf{V}}$ are the estimates of $\boldsymbol{\beta}$ and $\mathbf{V}$. Equation (8.10) is termed the "generalized least-squares solutions." Generally it is not practical to solve for $\boldsymbol{\beta}$ using these equations because they require the inverse of the matrix $\mathbf{V}$, which is of rank equal to the total number of observations and is not diagonal. Thus solutions are generally derived from the mixed model equations presented in Equation (6.16).

As noted earlier, the solutions for $\boldsymbol{\beta}$ are functions of the estimates of the VC. Differentiating Log L with respect to the variance requires the derivatives of $|\mathbf{V}|$ and $\mathbf{V}^{-1}$. For any square matrix, $\mathbf{M}$, the derivatives of $|\mathbf{M}|$ and $\mathbf{M}^{-1}$ with respect to a scalar x are computed as follows (Searle, 1982):

$$\frac{\partial(\ln|\mathbf{M}|)}{\partial x} = \text{tr}\left(\frac{\mathbf{M}^{-1}\partial\mathbf{M}}{\partial x}\right) \tag{8.11}$$

$$\frac{\partial\mathbf{M}^{-1}}{\partial x} = \mathbf{M}^{-1}\left(\frac{\partial\mathbf{M}}{\partial x}\right)\mathbf{M}^{-1} \tag{8.12}$$

where tr(.) is the trace of a matrix, computed as the sum of the diagonal elements. Using these equations, the derivative of Log L with respect to σ_i^2, the vector of VC is

$$\begin{aligned}\frac{\partial(\text{Log } L)}{\partial\sigma_i^2} &= -\frac{1}{2}\text{tr}\left(\mathbf{V}^{-1}\mathbf{V}_i\right) + \frac{1}{2}\left(\mathbf{y}-\mathbf{X}\hat{\boldsymbol{\beta}}\right)'\mathbf{V}^{-1}\mathbf{V}_i\mathbf{V}^{-1}\left(\mathbf{y}-\mathbf{X}\hat{\boldsymbol{\beta}}\right) \\ &\quad + \frac{1}{2}\left(\hat{\boldsymbol{\beta}}-\boldsymbol{\beta}\right)'\mathbf{X}'\mathbf{V}^{-1}\mathbf{V}_i\mathbf{V}^{-1}\mathbf{X}\left(\hat{\boldsymbol{\beta}}-\boldsymbol{\beta}\right)\end{aligned} \tag{8.13}$$

where $\mathbf{V}_i = \partial\mathbf{V}/\partial\sigma_i^2 = \mathbf{I}$ for $\sigma_i^2 = \sigma_e^2$ and $\mathbf{V}_i = \mathbf{Z}'\mathbf{A}\mathbf{Z}$ for $\sigma_i^2 = \sigma_u^2$. Setting these equations equal to zero, $\hat{\beta} = \boldsymbol{\beta}$, and rearranging, this equation becomes

$$\text{tr}\left(\hat{\mathbf{V}}^{-1}\mathbf{V}_i\right) = \left(\mathbf{y}-\mathbf{X}\hat{\boldsymbol{\beta}}\right)'\hat{\mathbf{V}}^{-1}\mathbf{V}_i\hat{\mathbf{V}}^{-1}\left(\mathbf{y}-\mathbf{X}\hat{\boldsymbol{\beta}}\right) \tag{8.14}$$

For $\sigma_i^2 = \sigma_e^2$ this equation becomes

$$\text{tr}\left(\hat{\mathbf{V}}^{-1}\right) = \left(\mathbf{y}-\mathbf{X}\hat{\boldsymbol{\beta}}\right)'\hat{\mathbf{V}}^{-1}\hat{\mathbf{V}}^{-1}\left(\mathbf{y}-\mathbf{X}\hat{\boldsymbol{\beta}}\right) \tag{8.15}$$

and for $\sigma_i^2 = \sigma_u^2$ the Equation (8.14) becomes

$$\text{tr}\left(\hat{\mathbf{V}}^{-1}\mathbf{ZAZ}'\right) = \left(\mathbf{y}-\mathbf{X}\hat{\boldsymbol{\beta}}\right)'\hat{\mathbf{V}}^{-1}\mathbf{ZAZ}'\hat{\mathbf{V}}^{-1}\left(\mathbf{y}-\mathbf{X}\hat{\boldsymbol{\beta}}\right) \tag{8.16}$$

The two equations are functions of both $\hat{\boldsymbol{\beta}}$ and $\hat{\mathbf{V}}^{-1}$, which appears on both sides of these equations. Furthermore, $\hat{\mathbf{V}}^{-1}$ is a nonlinear function of the VC. Thus iterative solutions are required to solve these nonlinear equations. The methods described in the previous chapter for iteration of nonlinear equations can be used.

Restricted Maximum Likelihood Estimation of Variance Components

The most common method for variance component estimation is restricted maximum likelihood (REML). REML differs from standard ML in that account is taken of the fact that the estimates of the fixed effects are not equal to their parameter values. The problem with standard ML estimation can be explained by considering the MLE of the variance for a normal distribution derived in Chapter 5. This estimate is $(1/N)\Sigma(y_i - \mu)^2$. Thus, the estimate of the variance is a function of μ, the actual mean, which is unknown. For standard estimation of variance from a sample, this problem is solved by replacing μ by the sample mean, and dividing by $N-1$, instead of N. Dividing by $N-1$ accounts for uncertainty in the value of the true mean. In mixed model variance component estimation, a parallel problem is encountered in that MLE assumes that the fixed effect solutions are equal to the true values.

In REML this problem is solved by a linear transformation of the observations that removes the fixed effects from the model. Consider the general mixed model. Define a matrix **K** such that $\mathbf{KX}=0$. Then

$$\mathbf{y}^* = \mathbf{Ky} = \mathbf{KZu} + \mathbf{Ke} \tag{8.17}$$

Searle *et al.* (1992) show that **K** satisfies the following relationship:

$$\mathbf{P} = \mathbf{K}'\left(\mathbf{KVK}'\right)^{-1}\mathbf{K} \tag{8.18}$$

where

$$\mathbf{P} = \mathbf{V}^{-1} - \mathbf{V}^{-1}\mathbf{X}\left(\mathbf{X}'\mathbf{V}^{-1}\mathbf{X}^{-1}\mathbf{XV}\right)^{-1} \tag{8.19}$$

so that

$$\mathbf{y}'^*\mathbf{V}^{*-1}\mathbf{y}^* = \left(\mathbf{y}'\mathbf{K}'\right)\left(\mathbf{KVK}'\right)^{-1}\mathbf{K}\left(\mathbf{Ky}\right) = \mathbf{y}'\mathbf{Py} \tag{8.20}$$

Substituting **Ky** for **y**, $\mathbf{KX}=0$ for **X**, **KZ** for **Z**, and $\mathbf{KVK}'$ for **V** in Equation (8.14) gives the following REML variance component estimators:

$$\mathrm{tr}\left(\hat{\mathbf{P}}\right) = \mathbf{y}'\hat{\mathbf{P}}\hat{\mathbf{P}}\mathbf{y} \tag{8.21}$$

for σ_e^2 and

$$\mathrm{tr}\left(\hat{\mathbf{P}}\mathbf{ZAZ}'\right) = \mathbf{y}'\hat{\mathbf{P}}\mathbf{ZAZ}'\hat{\mathbf{P}}\mathbf{y} \tag{8.22}$$

for σ_u^2. As in the case of ML, these nonlinear equations can only be solved by iteration. Estimates of REML VC can be obtained by derivative-based and derivative-free methods (see Lynch and Walsh, 1998, for a detailed explanation).

Estimation of Variance Components via the Gibbs Sampler

As noted in Chapter 5, GS is particularly well adapted to sampling the Bayesian posterior distribution, which is typically specified as a collection of conditional distributions. In the example given in the previous chapter for Bayesian estimation of the parameters of the granddaughter design, it was necessary to "integrate out" the random effects. This of course requires extensive computing.

Estimation of VC via ML or REML still generally requires solutions to the mixed model equations at each iteration, and this is the main computational limitation to the application of REML. As noted by Van Tassell *et al.* (1995), estimation of VC by GS has several advantages:

1. No solution to the mixed model equations is needed.
2. When simple sparse matrix techniques are used, analysis of data files larger than those using REML may be possible.
3. GS yields direct and exact estimates of VC and breeding values and confidence intervals for those estimates.

For estimation of VC in the mixed model, the joint density of interest is the distribution of fixed effects, random effects, and VC, all given the data. The marginal densities of interest are the distribution of fixed effects, random effects, and VC, given the data.

Van Tassell *et al.* (1995) assumed a simple mixed model with n records; only a single fixed effect, the general mean; and an additive genetic effect with r levels. Prior distributions are needed to describe the Bayesian model completely. For the fixed effects, Van Tassell *et al.* (1995) assumed a "flat" prior (i.e., the distribution was assumed to be proportional to a constant), indicating no prior knowledge about these effects. The random effects were assumed to be normally distributed with a variance of $\mathbf{A}\sigma_a^2$, and residuals were also assumed to be normally distributed with a variance of $\mathbf{I}\sigma_e^2$ where $\mathbf{A}$ is the numerator relationship matrix, as defined in Chapter 6, and the other terms are as defined previously.

Since VC can only have positive values, a continuous distribution with limits of zero and infinity should be used as the prior, for example, a chi-squared distribution. Van Tassell *et al.* (1995) used the inverted gamma (IG) distributions as the priors for the additive genetic and residual. The density function conditional on two parameters is as follows:

$$f\left(\sigma_i^2 | \alpha_i, \gamma_i\right) = \frac{i}{\Gamma(\alpha_i)\gamma_i^{\alpha_i}}\left(\sigma_i^2\right)^{-\alpha_i - 1} e^{-1/\gamma_i\sigma_i^2} \tag{8.23}$$

where i in σ_i^2 refers either to the additive genetic or residual variance, α_i is a shape parameter describing the uncertainty of the knowledge about variance component i, γ_i is the scale parameter that determines the expected value of variance component i, and $\Gamma(\alpha_i)$ is the gamma function for α_i, computed as follows:

$$\Gamma\left(\alpha_i\right) = \int_0^\infty t^{\alpha_i - 1} e^{-t} dt \tag{8.24}$$

The values of γ_i were chosen so that the expected value of the prior distributions was equal to the simulated value of the VC. $\alpha_i = 2.000001$ was assumed, so that the variance of the prior was finite and the distribution was as flat as possible. With $\alpha_i < 2$ the variance of the distribution is infinite. Thus even though the simulated values were used for γ_i, the prior knowledge was assumed to be very vague.

After some rather complicated algebra the conditional densities for the fixed and random effects and the additive genetic and residual VC were calculated as follows:

$$\boldsymbol{\beta} | \mathbf{u}, \sigma_e^2, \mathbf{y} \sim N\left(\hat{\boldsymbol{\beta}}, \left(\mathbf{X}'\mathbf{X}\right)^{-1} \sigma_e^2\right) \tag{8.25}$$

$$u_i | \beta, \mathbf{u}_{-i}, \sigma_a^2, \sigma_e^2, \mathbf{y} \sim N\left(\frac{1}{1 + a^{ii}\alpha}\left(y_i - \mathbf{x}_{(1)}\boldsymbol{\beta} - \mathbf{a}^{-1}\mathbf{u}_{-i}\alpha\right), \frac{\sigma_e^2}{1 + a^{ii}\alpha}\right) \tag{8.26}$$

for individuals with records, and

$$u_i|\mathbf{u}_{-i},\sigma_a^2,\sigma_e^2 \sim N\left(\frac{1}{1+a^{ii}}\left(\mathbf{a}^{-1}\mathbf{u}_{-i}\alpha\right),\frac{\sigma_e^2}{a^{ii}\alpha}\right) \tag{8.27}$$

for individuals without records, and

$$\sigma_a^2|\mathbf{u} \sim \mathrm{IG}\left(\frac{r}{2}+\alpha_a,\frac{1}{\frac{1}{2}\left(\mathbf{u}'\mathbf{A}^{-1}\mathbf{u}\right)+\frac{1}{\gamma_a}}\right) \tag{8.28}$$

$$\sigma_e^2|\boldsymbol{\beta},\mathbf{u},\mathbf{y} \sim \mathrm{IG}\left(\frac{n}{2}+\alpha_e,\frac{1}{\frac{1}{2}\left(\mathbf{y}-\mathbf{X}\boldsymbol{\beta}-\mathbf{Z}\mathbf{u}\right)'\left(\mathbf{y}-\mathbf{X}\boldsymbol{\beta}-\mathbf{Z}\mathbf{u}\right)+\frac{1}{\gamma_e}}\right) \tag{8.29}$$

where ~ $N(\mu,\sigma^2)$ indicates that the variable has a normal distribution with mean μ and variance σ^2, ~ IG(α, γ) indicates that the variable has an IG distribution with parameters α and γ, a^{ii} is the diagonal element i of $\mathbf{A}^{-1}$, $\mathbf{a}^{-1}$ is the row i of $\mathbf{A}^{-1}$ with a^{ii} removed, $\alpha=\sigma_e^2/\sigma_a^2$, $\mathbf{u}_{-i}$ is the vector of random animal effects without element i, and the other terms are as defined previously.

The GS algorithm can be described as follows:

1. Calculate μ as the arithmetic mean of the observations.
2. Calculate $u_i=h^2(y_i=\mu)$, $i=1, \ldots, n$, where $h^2=\sigma_a^2/\left(\sigma_a^2+\sigma_e^2\right)$.
3. Calculate $(y-\mathbf{X}\boldsymbol{\beta}-\mathbf{Zu})'(y-\mathbf{X}\boldsymbol{\beta}-\mathbf{Zu})=\Sigma(y_i-\mu-u_i)^2$.
4. Generate σ_e^2 from its IG density.
5. Generate μ from the density of $\boldsymbol{\beta}$.
6. Calculate $\mathbf{u}'\mathbf{A}^{-1}\mathbf{u}$.
7. Generate σ_a^2 from its IG density.
8. Calculate $\alpha=\sigma_e^2/\sigma_a^2$.
9. Generate u_i, from its normal density, for $i=1, 2, \ldots, n$.
10. Repeat steps 3 through 9.

After a "burn-in" of 100 rounds, 5000 iterations of GS loop were computed, and sums of squared from each 20th iteration were used to estimate the posterior density of the VC. To estimate the GS mean, the expected value of the VC, given the current value of the sum of squares and priors, was calculated and averaged. Based on the expected value of an IG variable:

$$E\left(\alpha_a^2|\mathrm{SSA}\right)=\frac{\frac{\mathrm{SSA}}{2}+\frac{1}{\gamma_s}}{\frac{r}{2}+\alpha_a-1} \tag{8.30}$$

and

$$E\left(\alpha_e^2 | \text{SSE}\right) = \frac{\dfrac{\text{SSE}}{2} + \dfrac{1}{\gamma_e}}{\dfrac{n}{2} + \alpha_e - 1} \tag{8.31}$$

where SSA and SSE are the additive genetic and residual sums of squares, respectively, and the other terms are as defined previously. The posterior density of the VC was calculated as the average of the conditional distribution of the VC.

In simulations of 400 animals, estimates of the means of the VC were unbiased, but estimates of the modes were biased for low heritabilities (Van Tassel *et al.*, 1995). By GS it is possible to compute confidence intervals for the VC estimates without approximations or assumptions of normality, based on the observed distribution of the results from the retained iterations.

Summary

Methods were presented to estimate the parameters for the mixed model equations, chiefly VC, by ML, REML, and GS. The main interest in simple ML estimates is historic. It is now clear that these estimates are biased, because they do not account for uncertainty with respect to estimates of the means of fixed effects. REML estimates are also biased, but less so, and yield estimates that must lie within the possible parameter space. This last property is by no means trivial. GS is the most computing intensive but has the advantage that it is possible to compute confidence intervals for the variance component estimates without approximations or assumptions of normality.

9 Distribution of Genetic Effects, Theory, and Results

Introduction

A number of studies have used stochastic simulations to evaluate marker-assisted selection (MAS) or genomic selection. Nearly all of these studies confronted the question of a mathematical model for the additive genetic variance that accounts for a finite number of QTL of sufficient magnitude for detection.

In this chapter we will first review the mathematical models that have been used to describe the polygenic variance, and then we will present formulas to calculate the effective number of QTL for a trait. In the next sections we will present results from very large human genomic studies which attempted to detect the individual QTL and show that detectable QTL explain only a small fraction of the total additive genetic variance. Recent studies show that the remainder of the genetic variance can be explained by many genes that are too small to identify even with very large studies. In the final sections we will compare these results to parallel studies with dairy cattle based on genetic linkage and population-wide linkage disequilibrium (LD).

Modeling the Polygenic Variance

As noted in the first chapter, the traditional mathematical model for polygenic variance has been the "infinitesimal model." That is, polygenic variance is assumed to be due to an infinite number of loci, each contributing an infinitesimal fraction of the total genetic variance. This model is mathematically tractable and apparently works very well, provided that no individual locus accounts for a very large fraction of the total genetic variance. However, the infinitesimal model cannot be applied to simulations of genomic selection models, which all postulate individual QTL large enough to be detected by linkage to genetic markers.

Nearly all of the simulation studies to be considered in this chapter, and a number of additional studies, have addressed the question of the appropriate mathematical model for polygenic variance with MAS. A number of studies have assumed a single segregating QTL on the background of the infinitesimal model for the remainder of the genetic variance (Gibson, 1994; Villanueva *et al.*, 1999). These simulations are not relevant to genomic selection, which in all cases assumes that many segregating QTL can be detected.

Genomic Selection in Animals, First Edition. Joel Ira Weller.

Most simulation studies have attempted to simulate all the additive genetic variance in terms of a finite number of loci, sampled from a theoretical distribution. Generally these studies have applied a distribution that postulates a few big QTL and many small ones. Zhang and Smith (1992, 1993) simulated a normal distribution of QTL effects, but they also considered a gamma distribution of QTL effects. Hayes and Goddard (2001), Meuwissen *et al.* (2001), and Weller *et al.* (2005) also assumed a gamma distribution of QTL effects. Some studies used a theoretical distribution to directly simulate the variance of each QTL, while other studies first simulated QTL effects and then either simulated allelic frequencies from a uniform distribution or assumed equal allelic frequencies. De Koning and Weller (1994) used a χ^2 distribution to simulate the variances of QTL effects. Hoeschele and VanRaden (1993a) postulated an exponential distribution, while Mackinnon and Georges (1998) assumed double exponential distribution of allelic effects at each QTL. The exponential distribution has the form

$$f(a) = \lambda e^{-\lambda a} \tag{9.1}$$

where a is the allelic effect and λ is the parameter of this distribution. The expectation of the distribution is equal to $1/\lambda$. They assumed that all QTL were biallelic and simulated allelic frequency from a uniform distribution. The double exponential distribution has the following form:

$$f(a) = \frac{1}{2}\lambda e^{-\lambda|a|} \tag{9.2}$$

Mackinnon and Georges (1998) simulated either two or four alleles for each QTL. The allelic frequencies were simulated by sampling from a uniform distribution and then dividing the sampled values by their sum, so that the sum of the allelic frequencies would equal unity. They assumed a heritability of 0.3, which was generated by simulating 5, 10, or 20 QTL. As the number of QTL increased from 5 to 20, it was necessary to increase λ from 6 to 12 to account for the total heritability of 0.3. With any of these theoretical distributions, there is no maximum value for QTL effect, although the probability of sampling a very large QTL becomes progressively smaller.

Hayes and Goddard (2001), Meuwissen *et al.* (2001), and Weller *et al.* (2005) assumed a gamma distribution for the distribution of QTL effects with scaling parameter α and shape parameter β. Defining x as the absolute difference between the substitution effects of the two paternal QTL alleles, $g(x)$, the distribution of x for each trait is

$$g(x) = \frac{\alpha^{\beta} x^{\beta-1} e^{-\alpha x}}{\int_0^{\infty} t^{\beta-1} e^{-t} dt} \tag{9.3}$$

The mode of the gamma distribution is $(\beta-1)/\alpha$. If $\beta<1$, the mode of the distribution will be at zero.

Lande and Thompson (1990) proposed the following deterministic distribution for the variances of the QTL:

$$\sigma_a^2(1-\alpha_a)\left[1, \alpha_a, \alpha_a^2, \alpha_a^3, \ldots\right] \tag{9.3}$$

The variances of the QTL generated by this model summed to infinity will equal σ_a^2. The parameter α, which must be between 0 and 1, determines the relative magnitude of the individual loci.

Assuming additivity, $\alpha_a = 2p(1-p)a$ for biallelic QTL, with a = the allelic substitution effect and p = allelic frequency. The first QTL in the series is the largest and has a variance of α_a. Subsequent QTL are progressively smaller, and the total number of loci is infinite. As α tends toward 0, the biggest QTL explains a relatively larger effect of the total additive genetic variance. With $\alpha_a = 0.5$, a single QTL explains half of the genetic variance. If the lth locus of series is the smallest QTL likely to be detected, then the maximum proportion of the additive genetic variance that can be detected is $1-a^l$.

The Effective Number of QTL

For the theoretical distributions considered in the previous section, the "effective number of loci" can be defined as the total additive genetic variance divided by the expectation of the individual QTL variances. Thus if QTL variances are generated by an exponential distribution, the effective number of loci will be equal to λ. Lande and Thompson (1990) defined a similar parameter for the distribution given in Equation (9.3):

$$N_{\mathrm{E}} = \frac{\left(\sum_{i=0}^{\infty} \alpha_a^{2i}\right)^2}{\sum_{i=0}^{\infty} \alpha_a^{2i}} = (1+\alpha_a)(1-\alpha_a) \tag{9.4}$$

where N_{E} is the effective number of loci. Values of α_a equal to 1, 1/3, 2/3, 5/6, and 11/12 correspond to N_{E} of 1, 2, 5, 11, and 23, respectively. As the effective number of loci increases, the fraction of the total additive genetic variance that can be detected, $1-a^l$, decreases.

The Case of the Missing Heritability

Numerous studies have shown that human height has a heritability of 0.8–0.9. Since this trait is highly heritable and easy to measure, has a nearly normal distribution, and has been recorded on very large samples, this should be an ideal trait to determine the fraction of variance that can be attributed to QTL detectable by genomic scans.

Visscher (2008) summarized the results of three research groups that analyzed hundreds of thousands of genetic markers genotyped on a total of 63,000 people measured for height. After correction for other factors, the effect of each SNP was tested on height with a linear model. A total of 54 QTL influencing this trait were validated. The validation stage is important because, when over 500,000 variants are tested, as was done in the current studies, many will be statistically "significant" by chance. (The problem of multiple comparisons will be considered in more detail in the following chapter.) The studies were based on an analysis of 14,000–34,000 individuals in the test stage and 6000–20,000 in the validation stages. The average effect size per "increasing" allele was approximately 0.4 cm, or approximately 0.8 cm between the two homozygous classes. No large effects explaining several centimeters were found.

Despite the huge sample sizes and huge numbers of markers analyzed, the sum of effects accounted for only 5% of the variance for height. Similar results were found for autism and schizophrenia, even though both diseases also have very high heritabilities. Maher (2008) gave several explanations for these disappointing results:

1. Markers used for whole genome analyses might not have been sufficiently dense to uncover most segregating QTL.
2. Heritability estimates may be inflated, but it is very unlikely that the "true" heritability may be much lower than the mean of many studies.
3. Most genetic factors affecting quantitative traits may be due to factors that are not "caught" by genome scans based on SNPs, such as copy number variation.
4. Some of the observed genetic effects may in fact be "epigenetic" effects—changes in gene expression that are inherited but not caused by changes in genetic sequence, for example, methylation of DNA.

The alternative explanation of classical quantitative genetics is that the naturally occurring variance in human height is due to a very large number of QTL, almost all of which have effects too small to be validated even by the very large sample sizes considered in these studies. For example, even with a sample size of 20,000 individuals, based on the stringent criterion of a type I error rate of 5×10^{-7} for the test of each SNP, the probability of detecting a QTL that explains 0.2% of the variance in two independent sample is only 0.81. For a sample half this size, the statistical power to detect a QTL of this size is reduced to 0.29 (Visscher, 2008).

Yang *et al.* (2010) estimated the proportion of variance for human height explained by 294,831 SNPs genotyped on 3925 unrelated individuals using a linear model analysis and validated the estimation method with simulations based on the observed genotype data. They showed that 45% of variance can be explained by considering all SNPs simultaneously. Thus, most of the heritability is not "missing" but has not previously been detected because the individual effects are too small to pass stringent significance tests. They provide evidence that the remaining heritability is due to incomplete LD between causal variants and genotyped SNPs, exacerbated by causal variants having lower minor allele frequency than the SNPs explored to date.

Their analysis was based on the construction of a "pseudo" relationship matrix based on the similarities among individuals for SNP genotypes considering all 294,831 SNPs. This matrix was then used in an REML analysis to estimate the fraction of the total variance for height explained by genetic relationships accounted for by the SNPs. A total of 0.45 of the phenotypic variance could be attributed to the SNPs, with a standard error of 0.08, which is 10-fold the variance explained considering only the SNPs with "significant" effects. Their results also suggest that the discrepancy between 80% heritability and 45% accounted for by all SNPs is due to incomplete LD between causal variants and the SNPs, possibly because the causal variants have lower minor allele frequencies on average than the SNPs typed on the array.

Methods for Determination of Causative Mutations for QTL in Animals and Humans

Since all the models described previously postulate the existence of a few large QTL, it is reasonable to ask if any of these have in fact been detected. It should first be noted that proving that a specific polymorphism is in fact the causative mutation, termed the quantitative trait nucleotide (QTN), is by no means trivial. Unlike major genes, individuals with a specific QTN genotype do not display a common phenotype. Glazier *et al.* (2002) noted that the most conclusive evidence that the

QTN has been found is a demonstration that replacement of the variant nucleotide results in swapping one phenotypic variant for another. Currently this is not possible for livestock species, but this situation may change in the near future with the development of CRISPR/Cas9 technology (https://www.neb.com/tools-and-resources/feature-articles/crispr-cas9-and-targeted-genome-editing-a-new-era-in-molecular-biology). Methods developed to find QTN suitable for plants and model animals cannot be applied to most livestock species, because of the lack of inbred lines, the long generation interval, the cost of each animal, and difficulty to produce transgenics or "knockouts."

As noted by Mackay (2001), "The only option…is to collect multiple pieces of evidence, no single one of which is convincing, but which together consistently point to a candidate gene." Methods to prove that a QTN has in fact been discovered were described in detail by Ron and Weller (2007). So far only four QTN that meet the criteria for QTN determination have been found for economic animal species: two in cattle, one in sheep, and one in swine.

Determination of QTN in Dairy Cattle

At least four QTN have been identified and verified by multiple studies in farm animals (Ron and Weller, 2007; Weller and Ron, 2011). Of these, two are in dairy cattle—*DGAT1* and *ABCG2* (Grisart *et al.*, 2002; Winter *et al.*, 2002; Cohen-Zinder *et al.*, 2005). The most significant effects for both genes are on fat and protein concentration, which have the highest heritabilities of all the traits routinely analyzed in dairy cattle. Both of these genes have effects on all the milk production traits. Unlike the situation for human height, each of these genes explains 50% of the variation in the trait with the greatest effect.

In 2002, two studies independently showed that a missense mutation, causing replacement of a lysine residue with alanine in exon VIII of the gene acyl-CoA:diacylglycerol acyltransferase (DGAT1) is the QTN (Grisart *et al.*, 2002; Winter *et al.*, 2002). Discovery was aided by the fact that DGAT1 was an obvious physiological candidate. In addition to mapping to the putative QTL region, DGAT1 encodes a microsomal enzyme that catalyzes the final step of triglyceride synthesis, and mice lacking both copies of DGAT1 are completely devoid of milk secretion. Complete concordance between this polymorphism and the QTL was found in three different dairy breeds.

The QTL near the middle of *Bos taurus* (BTA) chromosome 6 affecting protein concentration was first detected by Georges *et al.* (1995) in the US Holstein population. This QTL was then detected in several other Holstein populations including Finnish Ayrshire cattle (Velmala *et al.*, 1999) and Norwegian cattle (Olsen *et al.*, 2002). Ron *et al.* (2001) reduced the confidence interval to 4 cM centered on microsatellite BM143. Olsen *et al.* (2005) used physical mapping and combined linkage and LD mapping to determine that this QTL is located within a 420,000 bp region between the genes ABCG2 and LAP3.

In 2005, two research groups claimed to have found the QTN in two different genes. Schnabel *et al.* (2005) claimed that the QTN is located in a poly(A) sequence in the promoter region of the osteopontin gene, also denoted SPP1, while Cohen-Zinder *et al.* (2005) claimed that the QTN is a missense mutation in exon 14 of the ABCG2 gene. Both studies based their claim on gene function and concordance of bulls with known genotypes. Both genes are differentially expressed in the mammary gland during lactation, as compared to the liver. Furthermore, antisense SPP1 transgenic mice displayed abnormal mammary gland differentiation and milk secretion (Nemir *et al.*, 2000). More recent studies have provided additional evidence that the missense mutation in ABCG2 is in fact the causative mutation (e.g., Olsen *et al.*, 2007).

For both the QTL on BTA 6 and 14, the polymorphisms analyzed apparently do not account for the entire effect observed in these chromosomal regions (Bennewitz *et al.*, 2004a; Kuhn *et al.*, 2004; Cohen-Zinder *et al.*, 2005). The effect associated with the missense mutation in ABCG2 explains the entire effect observed on milk yield and fat and protein concentration, but does not explain the effects associated with fat and protein yield.

Many studies have found a QTL affecting all five milk production traits and somatic cell score (SCS) near the middle of BTA 20. Blott *et al.* (2003) claimed that a missense mutation in the bovine growth hormone receptor was responsible for the QTL affecting milk yield and composition on BTA 20, but did not find concordance for the bulls heterozygous for the QTL. Thus, this polymorphism may be responsible for only part of the observed effect on BTA 20 or may be a physiologically neutral mutation in LD with the QTN.

Estimating the Number of Segregating QTL Based on Linkage Mapping Studies

Hayes and Goddard (2001) proposed to estimate the total number of segregating loci in outbred populations, M, based on the following algorithm. One individual will only be heterozygous for fraction, $2K$, of the total number of genes segregating. Given the number of QTL found heterozygous in each sire N_i, an estimate of M is therefore $\bar{N}/2K$, where $\bar{N}$ is the mean of N_i. In order to calculate $2K$, Hayes and Goddard (2001) assumed the distribution of gene frequencies is for a population previously without selection for the quantitative trait, and with all genes neutral with respect to fitness. Clearly these assumptions are not correct, particularly if the population has undergone some artificial selection, in which case QTL with large effects are likely to be at extreme frequencies. In this case $2K$ will be underestimated.

Distribution of gene frequencies will reflect the generation of new alleles by mutation and their loss by drift. The gene frequency probability density from this assumption is $f(p)=K/(p(1-p))$ where p is the frequency of one allele. This calculates the gene frequency distribution accurately if the product of the mutation rate per locus and effective population size is small. The relevant part of the gene frequency distribution is from $\pi \le p \le 1-\pi$, where $\pi = 1/(2N_e)$.

The value of p is the lowest possible gene frequency in a population of effective size N_e. Again the effective population size as an approximation. All parents in the population are targets for mutation, but in modern livestock populations, mutations may only be exploited if they occur in elite breeding animals, as it is these animals which provide genetic material for the improvement of the population. The size of the elite population is likely to approximate the effective population size. The constant, K, is chosen so that

$$\int_{\pi}^{1-\pi} \frac{K}{p(1-p)} dp = 1 \tag{9.5}$$

Integrating this function and solving for K give the result:

$$K = \frac{1}{2\ln\left((1-\pi)/\pi\right)} \tag{9.6}$$

The mean heterozygosity of QTL is

$$\int_{\pi}^{1-\pi} \frac{2p(1-p)K}{p(1-p)} dp \tag{9.7}$$

which can be approximated as $2K$, or

$$\frac{1}{\ln\left((1-\pi)/\pi\right)} \tag{9.8}$$

This is the mean heterozygosity among the loci that are segregating and depends only on N_e. Hayes and Goddard (2001) calculated the total number of QTL segregating in the population with $N_e = 50$, 500, and 5000 ($\pi = 0.01$, 0.001, and 0.0001, respectively).

Pig data were from crossbreeding experiments between divergent breeds. Growth, carcass, and meat quality were analyzed. The three dairy experiments used a granddaughter design for QTL detection, with effects reported within grandsire families. Milk, fat, and protein yield and protein and fat percentage were analyzed. Overall, 50 significant QTL effects were reported.

For pigs a greater number of QTL within the range of 0.3–0.5σ_p, where σ_p is the phenotypic standard deviation, were detected than QTL > 0.5σ_p. The average QTL effect was $0.42\sigma_p \pm 0.02\sigma_p$. For the dairy cattle, the number of QTL < 0.3σ_p was larger than for the pig data set. Dairy experiments generally have more power to detect small QTL than the pig experiments. The average QTL effect for dairy data was $0.32\sigma_p \pm 0.03\sigma_p$. Both distributions were moderately leptokurtic, implying many QTL of small effect and few of large effect.

The number of segregating QTL for dairy cattle was predicted to be 49, 74, and 99 for $N_e = 50$, 500, and 5000, respectively. The world's dairy population is increasingly dominated by the Holstein breed. In addition, widespread use of artificial insemination has reduced the number of bulls used to breed dairy sires worldwide. As a result, the current N_e of the world's Holstein population is approximately 50 (Goddard, 1992), but this is a recent phenomenon, and N_e was much larger in the past.

Results of Genome Scans of Dairy Cattle by Granddaughter Designs

Genome scans by the granddaughter design have been completed for Holsteins from Canada (Nadesalingam *et al.*, 2001), the Netherlands (Spelman *et al.*, 1996; Schrooten *et al.*, 2000), France (Bennewitz *et al.*, 2003; Boichard *et al.*, 2003), Germany (Bennewitz *et al.*, 2003; Kuhn *et al.*, 2003), New Zealand (Spelman *et al.*, 1996), and the United States (Georges *et al.*, 1995; Ashwell *et al.*, 1996, 1997, 1998a, 1998b, 2001, 2004; Zhang *et al.*, 1998; Ashwell and Van Tassell, 1999; Heyen *et al.*, 1999); Finnish Ayrshires (Vilkki *et al.*, 1997; Viitala *et al.*, 2003; Schulman *et al.*, 2004); French Normande and Montbeliarde cattle (Boichard *et al.*, 2003); Norwegian cattle in Norway (Klungland *et al.*, 2001; Olsen *et al.*, 2002); and Swedish Red and White (SRB) (Holmberg and Andersson-Eklund, 2004). Daughter design analyses have been performed for Israeli Holsteins (Mosig *et al.*, 2001; Ron *et al.*, 2004).

Most studies have considered the five economic milk production traits—milk, fat, and protein production and fat and protein concentration—although a number of studies have also considered SCS, female fertility, herd life, calving traits, health traits, temperament, conformation, and other traits. The April 27, 2015, release of the Cattle QTLdb contains 17,908 QTLs from 588 publications based on marker–QTL linkage analyses (http://www.animalgenome.org/cgi-bin/QTLdb/BT/index). Those QTL represent 514 different traits, the overwhelming majority detected by granddaughter designs. Significant effects were found on all 29 autosomes, but most effects were found only in single studies and have not been repeated. Khatkar *et al.* (2004) performed a meta-analysis, combining data from most of these studies for milk, fat, and protein production, fat and protein concentration, and SCS which are summarized at (http://www.vetsci.usyd.edu.au/reprogen/QTL_Map/). They found significant across-study effects on chromosomes 1, 3, 6, 9, 10, 14, and 20.

Results of Genome-Wide Association Studies in Dairy Cattle by SNP Chips

The results of VanRaden *et al.* (2009) for the US Holstein population confirm that at least with respect to the QTN that have been detected, results of genome-wide association studies (GWAS) do correspond to the results of daughter and granddaughter designs. The largest effect found for fat concentration was located on BTA 14 flanking the *DGAT1* gene, with lesser effects on milk and fat yield. Similarly, the largest effect found for protein concentration was on BTA 6 flanking the *ABCG2* gene. The US analysis of 5285 bulls revealed few other large effects. Markers on BTA 18 had the largest effects on calving ease, several conformation traits, longevity, and total merit (Cole *et al.,* 2009).

Prediction accuracy was highest using a heavy-tailed prior assuming that each marker had an effect on each trait (Bayes A), rather than assuming a normal distribution of effects as in a linear model, or that only some loci have nonzero effects (Bayes B). (Bayes A and Bayes B models will be explained in detail in Chapter 10.) Results validate quantitative genetic assumptions that most traits are due to the contributions of a large number of genes of small additive effect, rather than support the "finite locus model," that is, only a small fraction of genes explain most of the genetic variance of quantitative traits.

Results of 912 Israeli Holstein bulls show that of the eight effects with the lowest probabilities for fat concentration under the null hypothesis of no effect associated with the marker, seven were located on BTA 14 in the vicinity of *DGAT1*. The SNP with the lowest probability was located at position 443,937 bp on BTA 14 and 1149 bp from the QTN (Grisart *et al.,* 2002). Similarly, the SNPs with the most significant effects on protein concentration formed a bracket from 37,024,132 to 37,454,409 bp on BTA 6. The *ABCG2* QTN is located within this bracket, 36,301 bp from its higher end (Cohen-Zinder *et al.,* 2005). In both cases GWAS identified segregating QTL within distances less than 40 kbp from the QTN.

More recently effects derived from genome scans for US Holsteins, Brown Swiss, and Jerseys for 33 traits—including milk, fat, and protein production, fat and protein percent, SCS, direct and maternal effect for stillbirth and calving ease, heifer and cow conception rate, productive life, net merit, and 19 conformation traits—are presented at https://www.cdcb.us/Report_Data/Marker_Effects/marker_effects.cfm?Breed=HO. For Holsteins, effects greater than 0.1 genetic standard deviations of the trait analyzed were found for fat production, fat percent, protein percent, productive life, heifer conception rate, sire calving ease and stillbirths, and teat length. No more than a single effect of this magnitude was found for any of the traits analyzed. Similar effects on the milk production traits were found for Jerseys, but not for the secondary traits. For Brown Swiss only one effect of this magnitude was found for the trait fore udder attachment, but the number of bulls analyzed was much smaller. Furthermore, no additional effects accounting for larger fractions of the genetic variances than the identified QTN were found for any of the traits analyzed.

Summary

Various theoretical models have been proposed to model the genetic variance as a function of a finite number of QTL. Among the statistical distributions proposed are the exponential, the double exponential, the chi-squared, and the gamma distribution. All of these studies assume that the additive genetic variance is due to a few rather large QTL and many small ones. Actual results from human height and disease traits demonstrate that only a very small fraction of the genetic variance can be explained by QTL with effects large enough to detect even if tens of thousands of individuals are genotyped. Based on these results several studies have suggested that a large fraction of the genetic variance is due to factors that cannot be detected by genome scans. More recent results with humans and cattle indicate that this is probably not the case. Rather, due to the multiple comparison problem, the majority of segregating QTL are just too small to be unequivocally detected even with very large sample sizes.

Although four QTN in animals were identified by 2007, virtually no progress has been made since then. This situation might change in the near future, due to huge reduction in costs for complete genome sequencing, the 1000 bull genomes project (http://www.1000bullgenomes.com/doco/hayes_pag_1000bullgenomes_2013.pdf), and new methods for QTN detection and determination, which will be discussed in detail in Chapter 20. In general results from the GWAS published to date validate quantitative genetic assumptions that most traits are due to the contributions of a large number of genes of small additive effect.

10 The Multiple Comparison Problem

Introduction

As we already noted in Chapter 1, it is now possible using DNA-level markers to obtain as many polymorphic markers as desired for any species of interest. The "multiple comparison problem" was noted briefly in the previous chapter. If the number of markers included in the analysis is large, two separate problems are encountered. First, the individual test type I error rate is no longer appropriate. For example, if 100 tests are performed, 5 should be "significant" at the 5% level purely by chance. The traditional approach to deal with multiple comparisons is to control the "family-wise (or experiment-wise) error rate" (FWER), instead of controlling the "nominal" or "comparison-wise error rate" (CWER). The FWER is controlled by setting the rejection threshold sufficiently strict, so that the probability that *any* of the null hypotheses tested are erroneously rejected is below a specified low level, usually 0.05. However, this severely reduces the power to detect true effects. Additional methods to deal with the problem of multiple comparisons will be considered.

A second problem with multiple marker analyses is that for those effects that are deemed "significant," the estimated effects will be biased upward (Georges *et al.*, 1995). The reason for this is that if the true effects are close to the critical value for significance, only those QTL with estimates greater than the true effects will meet the significance criterion. This problem will be considered in the last three sections of this chapter.

Multiple Markers and Whole Genome Scans

Lander and Botstein (1989) first considered the problem of multiple markers in detail. They presented analytical formula for two specific situations: a "sparse" map and a "dense" map. Kruglyak and Lander (1995) also considered the case of intermediate spacing. In the former they assumed that the markers were sufficiently far apart that the individual tests could be considered independent. In this case the FWER can be computed by the "Bonferroni correction" (Simes, 1986). The experiment-wise type I error rate is approximately equal to α_f/M, where α_f = FWER, α_c = CWER, and M = number of markers. For example, if 100 tests are performed and an FWER of 0.05 is desired, the CWER or nominal error rate must be approximately 0.0005.

Genomic Selection in Animals, First Edition. Joel Ira Weller.

This estimate can be marginally improved by observing that assuming statistical independence of the individual tests, the worst-case scenario, the type I error probability equals (Šidák, 1967)

$$\alpha_{\mathrm{f}} = 1 - \left(1 - \alpha_{\mathrm{c}}\right)^{M} \tag{10.1}$$

One can improve upon the simple Bonferroni/Šidák correction using the observation that the individual marker association tests are not statistically independent, but dependent to a degree which can be measured in terms of the pairwise haplotypic correlation between markers observed in the empirical data. An approach of this kind was used by Cheverud (2001) and Nyholt (2004), where an estimated effective number of tests was computed as follows:

$$M_{\mathrm{e}} = 1 + \left(M - 1\right)\left(1 - \frac{\mathrm{Var}(\lambda)}{M}\right) \tag{10.2}$$

where M_{e} = effective number of test and Var(λ) = the variance of the eigenvalues of the pair correlation matrix of test. M_{e} is then used instead of M in a Šidák-style correction. Other algorithms have all been proposed to estimate M_{e} (e.g., Moskvina and Schmidt, 2008).

In the dense map case, Lander and Botstein (1989) assumed that the markers are sufficiently close so that all "sites" along the chromosome are being tested for segregating QTL. In this case, the expected number of regions with a standard normal distribution value greater than the critical value for α_{c} under the null hypothesis of no segregating QTL anywhere in the genome, $\mu(Z)$, can be computed as follows for either the backcross (BC) or F-2 designs (Lander and Kruglyak, 1995):

$$\mu(Z) = \left[N_{\mathrm{c}} + 2\rho_{\mathrm{M}} M_{\mathrm{G}} Z^{2}\right]\alpha_{\mathrm{c}} \tag{10.3}$$

where N_{c} = number of chromosomes, ρ_{M} = the expected rate of recombination per Morgan, M_{G} = genome length in Morgans, and Z = the standard normal distribution value for α_{c}. For BC and half-sib designs, $\rho_{\mathrm{M}} = 1$, because recombination is followed only for a single chromosome. For a genomic scan of intermediate density, this equation can be modified as follows (Kruglyak and Lander, 1995):

$$\mu(Z) = \left[C + 2\rho_{\mathrm{M}} M_{\mathrm{G}} Z^{2} v\left(2Z\sqrt{\Delta}\right)\right]\alpha_{\mathrm{c}} \tag{10.4}$$

where Δ is the mean map distance between markers in Morgans and $v(2Z\sqrt{\Delta})$ represents a function of $2Z\sqrt{\Delta}$. For small values of Δ, $v(2Z\sqrt{\Delta})$ is approximately equal to $e^{-1.166Z\sqrt{\Delta}}$. For larger values of Δ, $v(2Z\sqrt{\Delta})$ is approximately equal to $1/(2Z^{2}\Delta)$. As Δ approached zero, the intermediate density function approaches the dense map function, and as Δ increases, this equation approaches $(N_{\mathrm{c}}+m)\alpha_{\mathrm{c}}$, which can be compared to the sparse map function of $\alpha_{\mathrm{f}} = m\alpha_{\mathrm{c}}$, given previously. The discrepancy between these two formulas is due to the fact that in Equation (10.4) it is assumed that each chromosomal interval is tested for a QTL, while the sparse map function of Equation (10.1) assumes that each marker is tested. Assuming that each chromosome has at least one marker, the number of chromosomal intervals including the chromosomal ends will be one more than the number of markers on each chromosome, or $N_{\mathrm{c}}+m$.

For small values of $\mu(Z)$, $\mu(Z)$ tends to α. This is because with low $\mu(Z)$ it is very unlikely that more than a single region can have a Z-value greater than the critical value. Lander and Botstein (1989) present a similar formula for likelihood ratio tests. For a dense map scan of the bovine genome by the daughter design ($C=30$ and $G=30$), a CWER of approximately 5×10^{-5}, comparable to a Z-value of 3.9, is required to obtain an FWER of 0.05. Requirement of such a stringent type I error results in a corresponding increase in the type II error. That is, many true effects will be missed.

Lander and Kruglyak (1995) proposed that, unless there is a reason to focus a priori on a specific chromosomal region, type I errors should be based on complete genome scans, even if the number of markers actually analyzed was limited. They maintain that even if the original marker spacing is quite wide, additional markers will be genotyped for those regions that display marginal significance. Thus, the whole genome is potentially under observation.

The problem of multiple comparisons is somewhat alleviated if those effects deemed "significant" are repeated on a second independent analysis. Since only these effects are considered in the second analysis, the number of comparisons is drastically reduced. However, this is not a viable option in most cases. For analysis of disease traits and generally for analysis of data on large animals, a second independent data set is not available. Two other methods that provide alternative solutions to the multiple comparison problem will now be considered.

QTL Detection by Permutation Tests

Churchill and Doerge (1994) proposed a method to empirically estimate FWER rejection thresholds that can be applied to a very wide range of experimental designs. Many different samples are generated from the actual data by "shuffling" the trait values with respect to the marker genotypes. Each individual genotyped is randomly assigned one of the trait values from the sample. Since the trait values for all individuals are now random with respect to marker genotypes, the null hypothesis of no linkage between the genetic markers and QTL is correct by definition. The test statistics computed from these "permutation samples" are then used to construct the empirical distribution of the test statistic under the null hypothesis. The appropriate rejection threshold for any desired comparison-wise or experiment-wise type I error can then be derived from the empirical distribution of the test statistic. This method has the advantage that no assumptions are required with respect to distributional properties of either the quantitative traits or the genetic markers. Rejection thresholds are computed based on the actual number and genomic distribution of markers genotyped. A disadvantage of this method is that thresholds must be computed anew by permutation for each data set analyzed.

Churchill and Doerge (1994) computed CWER and FWER based on permutation tests for simulated data. The fact that no assumptions are made with respect to the distribution of the test statistic under the null hypothesis is especially important for computation of the FWER. As demonstrated in the previous section, to obtain a reasonable FWER for a complete genome scan, a very small CWER is required. At these very low probabilities, it is likely that minuscule divergence of the actual data distribution from the theoretical distribution may result in a significant divergence of the analytically computed probability from the actual probability for the specific data set analyzed.

QTL Detection Based on the False Discovery Rate

Benjamini and Hochberg (1995) proposed controlling the "false discovery rate" (FDR) as an alternative to controlling the FWER for the general problem of multiple testing. They defined the FDR as "The expected proportion of true null hypotheses within the class of rejected null hypotheses."

Derivation of rejection thresholds based on controlling the FDR and important properties of this method will be described. We will then present examples based on actual data.

Assume that m multiple comparisons are tested. For each null hypothesis $H_1, H_2, \ldots, H_m$, a test statistic and the corresponding p-values, $P_1, P_2, \ldots, P_m$, are computed. Let $P_{(1)}, P_{(2)}, \ldots, P_{(m)}$ be the ordered p-values, and denote by $H_{(i)}$ the null hypothesis corresponding to $P_{(i)}$. If all null hypotheses are true, but K hypotheses, $H_{(1)}$ to $H_{(K)}$, are rejected, then the expectation of the number of hypotheses rejected should be approximately equal to the actual number of hypotheses rejected for any value of K. If, in fact, some of the null hypotheses are false (i.e., actual effects are detected), then the expectation of the number of hypothesis rejected should be less than K. The expectation of the number of hypotheses rejected assuming that all null hypotheses are true is $mP_{(K)}$. Defining $q=mP_{(i)}/i$, Benjamini and Hochberg (1995) proved that the FDR can be controlled at some level q^* by determining the largest i for which $q^*<mP_{(i)}/i$. That is, out of K hypothesis rejected, it is expected that the proportion of erroneously rejected hypotheses is no greater than q^*. Illustrative examples and important properties of the FDR will now be considered.

Weller *et al.* (1998) was the first application of the FDR to the QTL detection. This was the method used in nearly all genome scans based on an analysis of thousands of markers to determine that many segregating QTL could be detected. Comparison of FDR and FWER will be illustrated using the example of Weller *et al.* (1998) for a granddaughter design analysis of the US Holstein population. A total of 1555 sons of 18 US grandsires were genotyped for 128 microsatellites. Daughter yield deviations (DYD) were analyzed by the following linear model for seven economic traits:

$$Y_{ijk} = \mathrm{GS}_i + M_{ij} + e_{ijk} \tag{10.5}$$

where Y_{ijk} is the DYD (VanRaden and Wiggans, 1991) for kth son of the ith grandsire with paternal allele j, GS_i is the effect of the ith grandsire, M_{ij} is the effect of the jth marker allele, progeny of the ith grandsire and e_{ijk} is the random residual. For each marker–trait combination, an F-statistic was computed for the paternal marker allele effect nested within grandsire. Thus, 896 comparisons were tested.

The comparisons with the 10 smallest p-values are given in Table 10.1. Assuming uncorrelated tests, only two F-values have an FWER less than 0.05. Using Lander and Kruglyak's (1995) criterion of "suggestive linkage" (FWER < 0.5 for a complete genome scan), only four null hypotheses would be rejected. If all 10 hypotheses are rejected, q and thus FDR are still less than 0.25, even though the FWER = 0.811. Thus, seven or eight marker–trait combinations should represent "true" effects and can be expected to repeat on a second population sample. Unlike FWER, $q=mP_{(i)}/i$ is not monotonic. For example, as i increases from 5 to 6 and from 9 to 10, *q decreases*. A decrease in q occurs when the increase in successive probabilities is low.

Results for q, FWER, and CWER, computed as the individual F probabilities, up to $i=30$ are plotted in Figure 10.1. For $i>50$, q and FWER are very close, with both close to unity. For $i=10$, p is still less than 0.05. Thus in this case, the criteria of controlling the FDR at 0.5 and a CWER of 0.05 give similar results.

These results were compared to the p-values computed from a typical permutation of the same genotype data against the trait data. The permutation results are plotted in Figure 10.2. Since the relationship between the markers and traits after permutation is random by definition, no null hypotheses should be rejected, and FDR and FWER would be similar. For the lowest F probability, FWER was 0.45, and q was 0.31. Thus, one hypothesis would be rejected with FWER controlled at 0.5, but not with FDR controlled at any reasonable level. For i values greater than 5, the FWER is

Table 10.1 Estimation of FDR for granddaughter design results.

I	Trait	Chromosome	Marker	F-value	p-Value	Expectation[a]	FWER	q
1	Fat %	14	15	11.157	10^{-8}	10^{-5}	10^{-5}	10^{-5}
2	Fat %	3	1	5.295	0.00003	0.025	0.024	0.012
3	Fat yield	14	15	4.146	0.00009	0.077	0.074	0.026
4	Protein %	2	4	5.279	0.00042	0.378	0.315	0.094
5	Protein %	3	8	4.246	0.00091	0.818	0.559	0.163
6	SCS[b]	22	1	3.819	0.00101	0.907	0.596	0.151
7	SCS	22	2	4.590	0.00124	1.112	0.671	0.159
8	Fat %	3	8	3.880	0.00194	1.734	0.823	0.217
9	Milk	7	3	3.466	0.00231	2.068	0.874	0.230
10	SCS	23	1	4.218	0.00242	2.166	0.885	0.217

[a] Expectation for the number of hypothesis rejected under the null hypothesis.
[b] Somatic cell score.

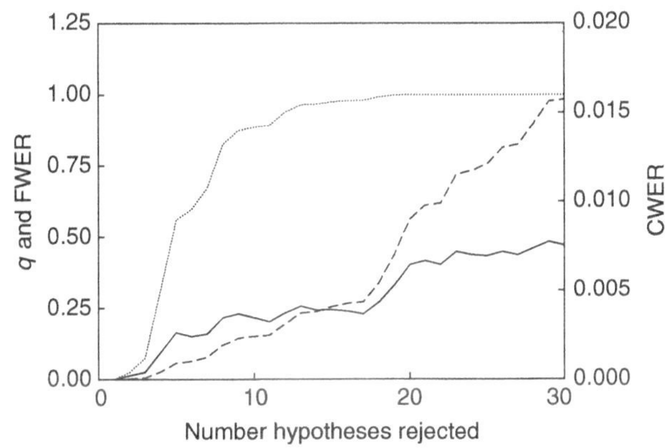

Figure 10.1 The q value (—), FWER (···), and comparison-wise type I error rate (CWER) (- - -) for the analysis of the granddaughter design data.

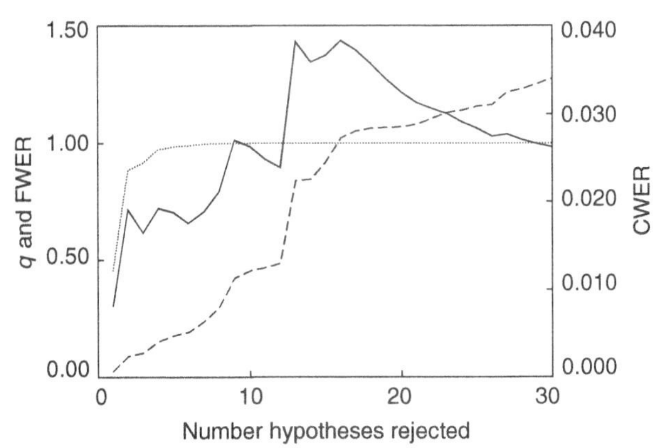

Figure 10.2 The q value (—), FWER (···), and comparison-wise type I error rate (CWER) (- - -) for the permuted granddaughter design data.

nearly equal to unity. By theory, the expectation of q is unity for all values of i, but this criterion is affected much more by random fluctuation than FWER. q is nearly equal to unity for $i=9$ but then rises to nearly 1.5 before settling down to close to unity by $i=30$. With $i=9$, CWER is still 0.01, which is almost exactly the expectation by chance (0.01×896 comparisons). Thus, by the criterion of CWER<0.01, 9 hypotheses would be rejected, as compared to 17 for the actual data. This illustrates the unreliability of the CWER criterion. The examples presented demonstrate the following important properties of the FDR:

1. If all null hypotheses are true, controlling FDR is equivalent to controlling FWER.
2. If some of the null hypotheses are false, then the FDR is smaller than the FWER. The difference between the two criteria increases with increase in the number of "false" null hypotheses (i.e., actual effects). Thus, any procedure that controls the FDR at a given level will also control the FWER at this level.
3. Unlike methods for controlling FWER, it is not necessary to assume that relationships among the test statistics are known. As demonstrated, the FDR can be readily controlled both for multiple linked markers and linked traits.
4. Even though $P_{(i)}$ increases monotonically with i, q does not. Thus, it may be necessary sometimes to *increase* i to control the FDR at the desired level.
5. Although the true FDR is less than q, as i increases, the FDR approaches q. This will be true even if the hypotheses are correlated.
6. By controlling the FDR, the number of hypotheses rejected, that is, QTL detected, is a function of the actual number of segregating QTL in the population; this is not true if either the FWER or CWER is controlled.
7. The dilemma of the appropriate rejection criterion for a partial genome scan is solved. The FDR can be controlled at the same level whether the complete genome or only part of the genome has been analyzed.
8. Additional levels of contrasts such as multiple traits or multiple populations can be handled without the necessity of a proportional increase in the critical test value.

A weakness of the FDR is that it tends to fluctuate widely for low i if the total number of hypotheses tested is very large.

A Priori Determination of the Proportion of False Positives

Controlling the FDR can only be applied after the experimental results are obtained. Thus it cannot be used to determine a priori the power of a planned experiment. Southey and Fernando (1998) and Fernando *et al.* (2004) proposed estimating the expected proportion of false-positive tests based on the assumed prior probabilities of true and false null hypotheses. For a single test, the expected proportion of false positives (PFP), $E(q)$, can be computed as follows:

$$E(q)=\frac{\alpha P(H_o)}{\alpha P(H_o)+P(H_\alpha)(1-\beta)} \tag{10.5}$$

where α is the nominal significance level (the type I error), $P(H_o)$ is the prior probability of the null hypothesis, $P(H_a)$ is the prior probability of the alternative hypothesis, and $(1-\beta)$ is the power of the test. If multiple tests are performed, then this equation becomes

$$E(q)=\frac{\sum\alpha_i P(H_{oi})}{\sum\alpha_i P(H_{oi})+P(H_{\alpha i})(1-\beta_i)} \tag{10.6}$$

where α_i is the significance level for test i, $P(H_{oi})$ is the prior probability for the null hypothesis for test i, $P(H_{\alpha i})$ is the prior probability for alternative hypothesis i, and $(1-\beta_i)$ is the power to reject this null hypothesis. If these probabilities are the same for all tests, then this equation reduces to

$$E(q)=\frac{m\alpha P(H_o)}{m\left[\alpha P(H_o)+P(H_\alpha)(1-\beta)\right]} \tag{10.7}$$

where m is the total number of tests. Fernando *et al.* (2004) denoted $E(q)$ the PFP. The problem with applying the PFP is that generally good estimates are not available for the prior probabilities. To compute these probabilities, Southey and Fernando (1998) assumed that the population was genotyped for N_k intervals of equal lengths and that N_Q QTL of equal size were scattered throughout the genome. They further assumed that there was no more than one QTL per interval. Thus the prior probability of the alternative hypotheses is N_Q/N_k, and the prior probability of null hypotheses is $1-N_Q/N_k$. They further assumed that these QTL explained all the genetic variance. With 10 QTL and heritability of 0.25, the probability of false positives was 0.3 with a significance level of 0.001 if 1000 individuals were genotyped in a BC design. This can be compared to the value of 5×10^{-5} required to obtain an FWER of 0.05 for a whole genome scan (Lander and Kruglyak, 1995).

Similar to the FDR, but unlike computation of the FWER, controlling the PFP is affected by the frequency of detectable QTL segregating in the population. However, unlike the FDR, the PFP is not affected by correlations among the tests even in extreme cases. Furthermore, this method can be used to plan an experiment in advance and answer the question: Is power sufficient to detect segregating QTL, provided they are present? However, in practice prior knowledge about the number, distribution, and effects of QTL is very vague. Similar to the Bayesian analysis of Hoeschele and VanRaden (1993a, 1993b) presented in Chapter 7, the assumptions made with respect to prior knowledge will affect the conclusions of the analysis.

Biases with Estimation of Multiple QTL

Smith and Simpson (1986) first noted that if multiple QTL are estimated as fixed effects, the estimated effects of those QTL that meet the "significance" criterion will be biased upward. This has been documented by simulation studies (Beavis, 1994; Georges *et al.*, 1995) and is supported by results of an actual experiment (Eshed and Zamir, 1996).

Georges *et al.* (1995) simulated a half-sib design but considered each family separately, so that the results are comparable to a BC design. The number of progeny in each family was varied from 50 to 200, and the QTL effects were varied from 0.25 to 1 phenotypic standard deviation. In all cases the QTL was bracketed by two markers 20 cM distant. The simulated QTL position was 5 cM from one of the markers. ML interval mapping was used to estimate QTL effect and location, and significance was determined by a likelihood ratio test. As simulated QTL effect or sample size decreased, the fraction of QTL determined as "significant" (LOD score ≥ 3) decreased, and bias of the estimated effect increased. Bias was under 10% of the simulated effect only if more than 90% of the simulated effects were "detected."

Beavis (1994) found an approximately linear relationship between the ratio of estimated to simulated QTL effect and the power of detection. If power of detection was only 10%, then the estimated effect was approximately fourfold the simulated effect. Furthermore, even if the simulated QTL were of equal size, the distribution of "significant" effects was positively skewed if power of detection was low.

Further support for these simulation studies comes from the results of Eshed and Zamir (1996). They analyzed the complete tomato genome for QTL affecting five quantitative traits using chromosomal segment substitution lines. The background parent was *Lycopersicon esculentum* (common tomato), and the donor parent was *Lycopersicon pennellii*. Fifty substitution lines, each containing a single chromosomal segment from *L. pennellii* on the background of the *L. esculentum* genome, were analyzed. Of 250 line-by-trait combinations, 81 were significantly different from the control isogenic line ($p<0.05$). The different substitution lines were then crossed to produce lines differing from the control each in two chromosomal segments. For those cases in which both *L. pennellii* chromosomal segments gave significant effects in the same direction, the effect estimates for the double substitution lines were consistently less than the sum of the effect estimates in the single chromosomal segment substitution lines. Eshed and Zamir (1996) proposed that these results were due to epistasis. However, this result is expected even without epistasis, if the "significant" effect estimates in the single segment substitution lines were overestimated. The effects should not be overestimated in the double segment analysis, because these effects are no longer a selected sample.

Bayesian Estimation of QTL from Whole Genome Scans: Theory

It should be possible to obtain unbiased estimates of a selected sample of effects if Bayesian estimation methods are used, as described in Chapter 7. In order to estimate QTL as random effects, it is necessary to know, or at least estimate, the variance of the distribution of effects. Methods to derive reasonable estimates for these parameters were considered by Hoeschele and VanRaden (1993a) and are summarized in Chapter 7. Actual information on the distribution of QTL effects was lacking prior to completion of whole genome scans.

Hayes and Goddard (2001) derived a mathematical distribution for QTL effects by combining results from several genome scans. Since the direction of the QTL effect relative to genetic markers is arbitrary, the QTL effect was assumed to be always positive. Weller *et al.* (2005) used data from a whole genome daughter design scan to estimate the prior distribution of QTL effects. Following Hayes and Goddard (2001), the QTL were assumed to follow a gamma distribution with scaling parameter α and shape parameter β. Hayes and Goddard (2001) assumed a common distribution for all traits, while Weller *et al.* (2005) derived a separate distribution for each trait analyzed.

Defining x as the absolute difference between the substitution effects of the two paternal QTL alleles, $g(x)$, the distribution of x for each trait is

$$g(x) = \frac{\alpha^{\beta} x^{\beta-1} e^{-\alpha x}}{\int_{0}^{\infty} t^{\beta-1} e^{-t} dt} \tag{10.8}$$

The mode of the gamma distribution is $(\beta-1)/\alpha$. If $\beta<1$, the mode of the distribution will be at zero. A normal distribution is assumed for the residuals of the observed effects. Thus the ordinate of observed QTL effect, $\hat{x}_i$, given the actual effect, $n(\hat{x}_i|x)$, will be

$$n(\hat{x}_i|x)=\frac{1}{\sqrt{2\pi\sigma_x^2}}e^{-\left((\hat{x}_i-x)^2/(2\sigma_x^2)\right)} \tag{10.9}$$

where σ_x = the standard error (SE) of the estimated QTL effect. This value will vary as a function of the experiment size.

As noted by Hayes and Goddard (2001), although the QTL effect is assumed always to be positive, the residual can be either positive or negative. Thus the density for $\hat{x}_i$, $f(\hat{x}_i)$, is computed as follows:

$$f(\hat{x}_i)=\int_0^\infty n(\hat{x}_i|x)g(x)dx+\int_0^\infty n(-\hat{x}_i|x)g(x)dx \tag{10.10}$$

The log-likelihood for the distribution of the QTL effects, Log $L(x)$, summed over all observed effects for each trait is

$$\text{Log}\,L(x)=\sum_{i=1}^{I}\text{Log}\left[f(\hat{x}_i)\right] \tag{10.11}$$

where I is the total number of estimated QTL effects per trait. Numerical integration was used to compute the density function, and Log $L(x)$ was maximized relative to α, β, and σ_x for each trait by a grid search for the three parameters. The prediction error variances of the parameter estimates were estimated by the negative of the inverse of the matrix of second derivatives of Log L at its maximum. The matrix of second derivatives was estimated numerically.

Bayes A and Bayes B Models

Meuwissen *et al.* (2001) distinguished between "Bayes A" models, which assume a continuous prior distribution of QTL effects with a nonzero effect for all comparisons tested, and "Bayes B" model, in which a zero effect is assumed for the majority of the comparisons tested. Thus the model presented earlier can be considered a Bayes A model. The following Bayes B model, which considers the possibility that only a fraction of the marker contrasts were associated with segregating QTL, was also tested:

$$f(\hat{x}_i)=P\left(\int_0^\infty n(\hat{x}_i|x)g(x)dx+\int_0^\infty n(-\hat{x}_i|x)g(x)dx\right)+2(1-P)\left(\int_0^\infty n(\hat{x}_i|x)dx\right) \tag{10.12}$$

where P = the fraction of marker contrasts associated with segregating QTL and the other terms are as defined previously.

The likelihood of the individual QTL effects given the distribution of QTL effects for a given trait, $L(y)$, was computed under the assumption that QTL genotype has been determined for each individual. For the daughter design, only the paternal allele is considered, and the progeny will be divided into two groups: the J_1 individuals that received the positive paternal QTL allele and the J_2 individuals that received the negative paternal allele. $L(y)$ is then computed as follows:

$$L(y) = g(x_i|\alpha,\beta)\prod_{j=1}^{J_1}\left(\frac{1}{\sqrt{2\pi\sigma_y^2}}e^{-\left(\frac{(y_j-0.5x_i)^2}{2\sigma_y^2}\right)}\right)\prod_{j=J_1+1}^{J}\left(\frac{1}{\sqrt{2\pi\sigma_y^2}}e^{-\left(\frac{(y_j+0.5x_i)^2}{2\sigma_y^2}\right)}\right) \quad (10.13)$$

where x_i is the effect for QTL i, y_j is standardized record of individual j, $J = J_1 + J_2$ is the total number of individuals genotyped for the QTL, σ_y^2 is the residual variance of the individual records, and the other terms are as described previously. Since the observations were normalized by subtraction of the mean of the two means, it is not necessary to include a mean effect in the likelihood. Since α and β are assumed known, this likelihood was maximized only relative to x_i and σ_y.

Log $L(y)$, the log-likelihood, with terms including only constants deleted is computed as follows:

$$\begin{aligned}\text{Log}\,L(y) &= (\beta-1)\text{Log}(x_i) - ax_i - J\left(\text{Log}(\sigma_y)\right)\\ &-\left[\frac{1}{2\sigma_y^2}\right]\left[\sum_{j=1}^{J_1}(y_j-0.5x_i)^2 + \sum_{j=J_1+1}^{J}(y_j+0.5x_i)^2\right]\end{aligned} \quad (10.14)$$

Solutions were obtained by a one-dimensional search with respect to x_i. At each value of x_i, the ML value for σ_y was determined by solving for $\partial[\text{Log}\,L(y)]/\partial\sigma_y = 0$ as follows:

$$\sigma_y^2 = \frac{\sum_{j=1}^{J_1}(y_j-0.5x_i)^2 + \sum_{j=J_1+1}^{J}(y_j+0.5x_i)^2}{J} \quad (10.15)$$

In the interval mapping QTL analyses, the genotype of each individual with respect to the QTL is not known with certainty. In this case $L(y)$ is computed as follows:

$$L(y) = g(x_i|\alpha,\beta)\prod_{j=1}^{J}\left[\left(\frac{p_j}{\sqrt{2\pi\sigma_y^2}}e^{-\left(\frac{(y_j-0.5x_i)^2}{2\sigma_y^2}\right)}\right)+\left(\frac{1-p_j}{\sqrt{2\pi\sigma_y^2}}e^{-\left(\frac{(y_j+0.5x_i)^2}{2\sigma_y^2}\right)}\right)\right] \quad (10.16)$$

where p_j is the probability that the progeny received the positive paternal allele, given its marker genotype. This likelihood was solved for x_i and σ_y by a two-dimensional grid search.

The prediction error variances of the QTL estimates were estimated by two methods. First, from the inverse of the two-by-two matrix of second derivatives for x_i and σ_y, as described for the parameters of the gamma distribution, these were denoted "empirical" values because the second derivatives were derived numerically. The second method applied the assumption that the minor diagonal elements of the matrix of second derivatives of Log $L(y)$ are small relative to the major diagonal elements. Under this assumption, $1/[\partial^2[\text{Log } L(y)]/\partial x_i^2]$ will be approximately equal to the prediction error variance of x_i. If the QTL genotype was assumed known without error, $\partial^2[\text{Log } L(y)]/\partial x_i^2$ is computed as follows:

$$\frac{\partial^2\left[\text{Log } L(y)\right]}{\partial x_i^2} = \frac{1-\beta}{x_i^2} - \frac{J}{4\sigma_y^2} \tag{10.17}$$

For large values of x_i, the first term on the right-hand side of this equation tends to zero. Similarly, $\partial^2[\text{Log } L(y)]/\partial\sigma_y^2$ can be derived by differentiating Equation (10.14) twice and substituting from Equation (10.15) as follows:

$$\frac{\partial^2\left[\text{Log } L(y)\right]}{\partial \sigma_y^2} = -\frac{2J}{\sigma_y^2} \tag{10.18}$$

which is the value of $\partial^2[\text{Log } L(y)]/\partial\sigma_y^2$ for a sample from a normal distribution. For both methods, SE estimates were derived as the square roots of the corresponding prediction error variances.

Bayesian Estimation of QTL from Whole Genome Scans: Simulation Results

The method was evaluated on a simulated daughter design genome scan with 1000 contrasts under the assumption that the true effects were sampled from a gamma distribution with α and β values equal to 1.99 and 0.90, respectively. For each contrast, an effect was simulated by random sampling from this gamma distribution, and a sample of 400 individual records was generated. Each individual had a 50% chance to receive the positive or the negative QTL allele. A random residual was generated by sampling from a normal distribution with mean zero and a standard deviation of 10. Thus the expected SE for the QTL effect for a balanced sample of 400 individuals will be equal to unity. The trait value for each individual was then computed as the residual +½ the QTL effect for individuals that received the positive allele and −½ the QTL effect for individuals that received the negative allele.

The least squares (LS) QTL effect was then estimated for each simulated QTL based only on the genotypes and trait records. If the absolute value of the t-value was greater than 2.5 (a probability of 0.012 for comparison-wise significance), then the QTL effect was also estimated by the Bayesian method, with the QTL genotypes assumed known.

There were 54 contrasts with t-values greater than 2.5, as compared to $1000 * 0.012 = 12$ expected purely by chance. Thus the FDR=0.22. The LS estimates were highly biased, with a mean value of 3.04, as compared to 1.45 for the simulated values. The mean of the ML estimates was 1.26, which is much closer to the simulated values. The standard deviation of the ML estimates was slightly higher than the LS estimates, although both standard deviations were considerably lower than the standard deviation of the simulated effects. The LS and Bayesian estimates for σ_y were both very close to the simulated value of 10. The R^2 of the simulated values was more than fivefold for the Bayesian estimates, as compared to the LS estimates, but both were less than 0.1.

Summary

With multiple markers, and the possibility of complete genome scans, comparison-wise type I error rates for individual tests are virtually meaningless. Furthermore, estimates of QTL effects deemed "significant" would be biased. Four methods were presented to deal with the problem of multiple comparisons: computation of error rates for complete genome scans, permutation tests, controlling the FDR, and a Bayesian analysis based on prior information on the distribution of segregating QTL in the population. None of these methods completely solve the problem of multiple comparisons. Various solutions have been presented to analyze multiple pedigrees, covering the range from a separate analysis of each family to a joint analysis with the same allele segregating in all families, but again there is no uniformly "best" solution. In the last three sections, we described Bayesian methods to deal with bias in the estimation of QTL effects due to "selection" of the significant effects. These methods are computing intensive and require assumptions with respect to the distribution of QTL effects in the population.

11 Linkage Mapping of QTL

Introduction

As first demonstrated in 1923 by Sax, a simple linear model can be used to detect linkage between segregating QTL and genetic markers in crosses between inbred lines. However, this analysis cannot be used to accurately map QTL or to derive unbiased estimates of QTL effects. These objectives can be obtained by a technique denoted "interval mapping."

Although the major thrust of this book will be estimation of QTL effects based on linkage disequilibrium (LD) mapping, LD mapping was derived based on interval mapping, which was first proposed in 1989 based on maximum likelihood (ML) methodology for crosses between inbred lines (Lander and Botstein, 1989). In 1992 methods were developed for application of linkage mapping via nonlinear regression. These methods were easier to apply than ML. Furthermore, by minor modifications it was also possible to apply these methods to daughter and granddaughter designs.

Interval Mapping by Nonlinear Regression: The Backcross Design

The nonlinear least squares method of QTL parameter estimation with two flanking markers was developed independently in 1992 by Haley and Knott and by Martinez and Curnow. We will illustrate the method using the backcross (BC) design, although the method has been adapted to most of the designs considered in the previous chapter with flanking markers. The BC design with two flanking markers is illustrated in Figure 11.1. For the BC progeny only the chromosome from the F-1 parent is shown. There are eight possible gametic haplotypes (including the QTL): two nonrecombinants, four single recombinants, and two double recombinants. The following model can be defined:

$$Y_{ij} = \mu_1\left(1 - p_i\right) + \mu_2 p_i + e_{ij} \tag{11.1}$$

where Y_{ij} is the production record of the jth individual with marker genotype i, μ_1 is the mean for individuals with genotype Q_1Q_2, μ_2 is the mean for individuals with genotype Q_2Q_2, p_i is the

Genomic Selection in Animals, First Edition. Joel Ira Weller.

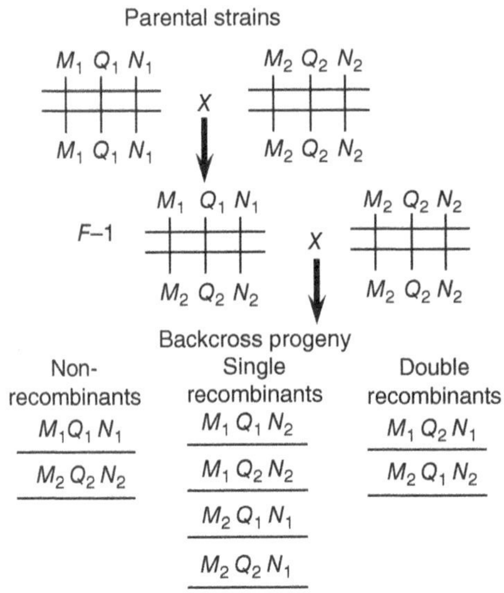

Figure 11.1 The backcross design with flanking markers.

probability that an individual with marker genotype i has genotype Q_2Q_2, and e_{ij} is the residual. This model can be simplified as follows:

$$Y_{ij} = \mu_1\left(\mu_2 - \mu_1\right)p_i + e_{ij} \tag{11.2}$$

p_i is a function of the recombination parameters and can be estimated for each of the four marker haplotypes—M_1N_1, M_1N_2, M_2N_1, and M_2N_2—as follows:

$$p_{M_1N_1} = \frac{r_1r_2}{1-R} \tag{11.3}$$

$$p_{M_1N_2} = \frac{r_1\left(1-r_2\right)}{R} \tag{11.4}$$

$$p_{M_2N_1} = \frac{r_2\left(1-r_1\right)}{R} \tag{11.5}$$

$$p_{M_2N_2} = \frac{1-r_1r_2}{1-R} \tag{11.6}$$

where R is the recombination frequency between the two markers M and N, r_1 is the recombination frequency between M and Q, and r_2 is the recombination frequency between Q and N.

If r_1 was known, it would be possible to substitute these values into Equation (11.2) and then solve as a simple linear regression, with μ_1 as the y-intercept and $\mu_2-\mu_1$ as the slope. Since r_1 is not known, Equation (11.2) can be considered as four separate equations, one for each marker haplotype.

Assuming that R is known without an error, it is possible to solve for r_2 in terms of R and r_1 for the assumed map function. For example, for the Haldane function (Haldane, 1919), which assumes zero interference,

$$R = r_1 + r_2 - 2r_1r_2 \tag{11.7}$$

$$r_2 = \frac{R - r_1}{1 - 2r_1} \tag{11.8}$$

This still leaves us with four equations, which are nonlinear functions of the QTL means and r_1. The least squares solution for this model, which is nonlinear in r_1 for all three parameters, will be the values that minimize the residual sum of squares as a function of RSS(r_1), computed as follows:

$$\mathrm{RSS}(r_1) - \sum_{i=1}^{4}\sum_{j=1}^{n_i}\left[Y_{ij} - \hat{Y}_{ij}(r_1)\right]^2 \tag{11.9}$$

where $\hat{Y}_{ij}(r_1)$ is the estimated value of Y_{ij} with recombination frequency of r_1 between the QTL and the first marker and n_j is the number of individuals in marker class i.

The least squares solutions can be derived by a nonlinear least squares algorithm, such as PROC NLIN of SAS (SAS Institute Inc., 1999). The appropriate ratio for the F-test is the model mean squares divided by the residual mean squares. The model sum of squares can then be computed as the total sum of squares less the residual sum of squares. In theory, the mean squares are derived by dividing the sums of squares by their degrees of freedom. Under the null hypothesis of no segregating QTL, this ratio should have an approximate central F-distribution.

Interval Mapping for Daughter and Granddaughter Designs

For the granddaughter design, several studies have suggested analyzing either estimate breeding values (EBV) (Andersson-Elkund *et al.*, 1990; Cowan *et al.*, 1992) or daughter yield deviations (DYD) (Hoeschele and VanRaden, 1993b) based on mixed models that include repeated records and fixed nuisance effects. As explained in section "Important Properties of Mixed Model Solutions" of Chapter 6, DYD are the daughter record means of each son adjusted for systematic environmental effects and merits of mates (VanRaden and Wiggans, 1991). The EBV or DYD is then analyzed by a linear model including only the effects associated directly with the genetic markers. EBV derived from a mixed model will be regressed toward the mean, and therefore estimates of QTL effects derived as described will be biased. In addition, the variances of either EBV or DYD will be a function of the quantity of information on the son. Thus, these studies have proposed to weight the evaluations by some function of their reliabilities, the coefficient of determination between the genetic evaluation and the actual genetic value. In the mixed model equations the coefficient matrix is multiplied by the inverse of the residual variance matrix. Therefore, for DYD, for which the variance decreases as the number of daughters increases, weighting by the reliabilities is approximately correct. However, for EBV derived from mixed model analyses, variances increase as the number of progeny increases. Therefore the effect of weighting by the reliability has an effect opposite to that desired. Sons with few daughters are multiplied by a smaller factor, even though their variance is less.

An additional problem in the analysis of segregating families is the question of whether different families should be analyzed jointly or separately. In crosses between inbred lines, if each individual phenotyped is also genotyped, then the polygenic effect of each individual is completely confounded with the other factors that make up the random residual associated with each individual. This is not the case with the daughter design where all daughters of a sire have a common polygenic effect. The common polygenic effect will not affect QTL genotype estimates computed within a family. The common polygenic effects can then be considered part of the general mean.

Knott *et al.* (1994, 1996) proposed that the nonlinear regression method could be used to estimate QTL effects for multiple pedigrees. They assumed that only a single QTL was segregating in the chromosomal segment under analysis, but did not make assumptions with respect to the number of different alleles present in the population. Based on these assumptions, they assumed that QTL location was the same for all families but estimated a separate QTL substitution effect for each family. This model is amenable to analysis by nonlinear regression and does not require estimation of a common family polygenic effect. The disadvantage is that all families are assumed to be heterozygous for the QTL, which will generally not be the case.

The analysis model is as follows:

$$Y_{ijk} = \mu_{1i}\left(1 - p_{ij}\right) + \mu_{2i}p_{ij} + e_{ijk} \tag{11.10}$$

where Y_{ijk} is the trait record for individual k of family i with marker genotype j, μ_{1i} and μ_{2i} are the means for progeny that received paternal QTL alleles 1 and 2 in family i, p_{ij} is the probability that a progeny of sire i with marker genotype j received paternal QTL allele 1, and e_{ijk} is the random residual. Although QTL location is assumed to be the same in all families, p_{ij} must be computed separately for each individual, because it will depend on which markers are informative in each progeny of each family. A marker will be "informative" only if the genotype of the progeny is different from the genotype of the sire. Otherwise it is not possible to determine which paternal allele was passed to the progeny. As noted also by Martinez and Curnow (1994), even if all markers to one side of the putative QTL location are uninformative in a specific individual, p_{ij} can still be calculated based on the recombination frequency between the assumed QTL position and the single linked marker. Thus, only individuals without any markers in linkage to the putative QTL location will be discarded from the analysis.

There are two main advantages of this method. First, data across families is combined to estimate the QTL location. This is especially important in daughter and granddaughter designs, because only some of the markers analyzed will be informative in each pedigree. Second, since an individual substitution effect is computed within each family, it is not necessary to estimate a common polygenic effect for each family. The main disadvantage of this method, as noted previously, is that each family is considered to be heterozygous for two different QTL alleles.

Computation of Confidence Intervals

For maximum likelihood estimation (MLE), the estimation error variance–covariance matrix can be estimated from the inverse of the ML matrix of second differentials. This is also the case for linear model estimation. The prediction error variance estimates can then be used to derive confidence intervals (CI) for all the parameters. This is not an option for interval mapping by the nonlinear regression method. Even for MLE this method of deriving CI has limitations. First, in some cases, the likelihood function cannot be readily differentiated twice for all parameters, especially if multiple markers and QTL are included in the analysis. Second, estimation of CI by a linear function of the square roots of the prediction error variances assumes that the distributions of the parameter estimates are symmetric. This of course will not be the case for variances, which can only be positive, but will also not be the case for recombination parameters, especially if the putative QTL location is close to a marker or the end of the chromosome. Alternative methods to estimate CI, especially for QTL location, have also been proposed.

Lander and Botstein (1989) proposed estimating "support intervals" for QTL location, based on the likelihood ratio test. In a likelihood ratio test, the likelihood maximized over all parameters is compared with the ML obtained with some of the parameters fixed. If the null hypothesis is correct, the log of the ratio of the two likelihoods times 2 should have a χ^2 distribution with degrees of freedom equal to the number of parameters fixed in the null hypothesis that are allowed to "float" in the alternative hypothesis. Similarly, the lower bound of the CI of $1-\alpha$ probability for any of the parameter estimates can be constructed based on the following statistic:

$$\chi^2_{(1-\alpha/2)} = 2\ln\left[\frac{L_{\max}}{L_{(\theta=\theta_0)}}\right] \quad (11.11)$$

where $\chi^2_{(1-\alpha/2)}$ is the χ^2 squared value for $1-\alpha/2$ with one degree of freedom, $L_{\max}$ is the likelihood value with the likelihood maximized over all parameters, and $L_{(\theta=\theta_0)}$ is the likelihood maximized over all parameters with θ fixed at θ_0, which is a value for the parameter θ less than the ML value, but closest to its ML value that gives the appropriate χ^2 value. Similarly the upper bound of the CI is determined by the same statistic with θ_0 computed as a value of θ greater than the ML value that satisfies this equation.

Mangin *et al.* (1994) showed that for QTL location, this support interval underestimated the actual CI, especially for small QTL effects. They were able to derive a rather complicated test statistic that accurately estimates the CI for small QTL effects, but the distribution of this test statistic must be computed empirically. Furthermore, this method does not account the possibility that the QTL is outside the marker bracket. In this case there is still likely to be a maximum for QTL location within the marker bracket (Martinez and Curnow, 1992). It does account for the possibility that the CI is asymmetric, which will generally be the case, especially if the QTL is located near an end of the chromosome.

Simulation Studies of CIs

To obtain accurate estimates of the CI by simulation, it is necessary to generate a large number of samples. For example, if 1000 samples are generated, the 95% confidence limits are obtained by determining the 25 lowest and 25 highest estimates for each parameter. Thus the effective number of samples can be considered to be 50. In much smaller samples, the estimated confidence limits will vary widely.

Darvasi *et al.* (1993) estimated QTL parameter estimation error variances based on the inverse of the matrix of second differentials and by repeated simulation for the BC design with marker brackets. The 95% CI was then estimated as ±2 estimation SE for each parameter. They also directly estimated the 95% CI for each parameter by repeated simulation. All methods were very accurate for estimation of QTL effect variances. Estimates based on the second differential matrix tended to slightly overestimate SE for QTL means relative to the empirical estimates, especially for large spacing between markers. Neither the QTL effect nor marker spacing had any appreciable effect on CI for QTL means. The effect of sample size was quadratic, as expected. That is, doubling the sample decreased the CI by a factor of about the square root of two.

For QTL map location, the estimates based on the empirical 95% CI and four times the empirical standard error were generally similar. However, estimates based on the second differential matrix tended to underestimate the CI for small marker intervals and overestimate the CI for large marker intervals. Differences were in some cases more than twofold. Clearly, for this parameter the asymptotic properties of the second differential matrix do not hold. For the BC design and a single marker, the matrix of second differentials tended to overestimate error variance for all parameters, even though by theory the opposite should occur. It should be noted though that even for very large samples, the error variance estimated by the matrix of second differentials is correct only at the point of ML. The likelihood function can behave marked differently for other parameter values.

Mackinnon and Weller (1995) estimated parameter SE both empirically and by the matrix of second differentials for the daughter design for a single marker and also analytically computed the 95% CI, as described earlier. In addition to QTL means, recombination rate, and the residual variance, they also estimated the QTL allele frequencies. CI estimates based on assuming that all other parameters were fixed tended to underestimate the SE derived by either repeat simulation or the matrix of second differentials. As for the BC design with a single marker, the matrix of second differentials tended to overestimate the SE, even though the opposite was expected. Discrepancies increased with decrease in sample size. CIs were largest for recombination rate. The standard error for r with a substitution effect of 0.5 was about 0.1 with 2000 individuals. For the BC design and a marker bracket of 50 cM, a similar SE was obtained with only 1000 individuals, although, in both cases, the number of QTL genotypes performed was the same.

Empirical Methods to Estimate CIs, Parametric and Nonparametric Bootstrap, and Jackknife Methods

In the "parametric bootstrap" method, parameter estimates are first derived by any of the methods considered. In the second step a large number of sample distributions of equal size to the actual data sample are then derived from the assumed theoretical distribution, assuming that the original parameter estimates are the parameter values. Parameter estimates are then derived for each sample. The CI for each parameter is then derived from the empirical distributions of the parameter estimates from the samples generated. The weakness of the parametric bootstrap method is that it assumes that both the theoretical distribution and the original parameter estimates are correct. If either of these assumptions is incorrect, then estimated CI can differ widely from the true values.

Efron and Tibshirani (1993) proposed empirical "bootstrap" methods to estimate CI in situations where analytical methods cannot be applied. In "nonparametric bootstrapping," a large number of repeat samples of size equal to the actual data are generated by sampling *with repeats* from the original data. Thus, in a particular sample some of the actual records will appear more than once, while other observations will be missing. If the actual data consists of at least several hundred observations, it will be possible to draw a virtually unlimited number of different samples in this method. The parameter estimates are then derived for each sample, and, as in parametric bootstrapping, the distribution of these estimates is used to derive empirical CI limits. This method is not strictly "nonparametric," because assumptions about the distribution are still employed to derive parameter estimates for each sample. This method is more robust to violations of assumptions used to derive parameter estimates.

"Jackknife" samples are derived from the original data sample by generating new samples consisting of the original data, with one observation deleted. Thus, unlike the empirical bootstrap, the number of jackknife samples that can be derived is only equal to the sample size. Bootstrap and jackknife sampling can be combined to analyze complex problems.

Visscher *et al.* (1996) applied the nonparametric bootstrap method to estimate CI for QTL location in a BC design with multiple markers and a single QTL segregating on the chromosome. Accuracy of the CI estimate was determined by the proportion of CIs that actually contained the QTL. They found that this method was able to estimate accurately the CI for QTL location, provided that the CI was less than two-thirds of the entire chromosome. If the CI estimate was larger than two-thirds of the chromosome, it tended to overestimate the actual CI. This is inevitable as the QTL effect and sample size become smaller. The estimated CI for QTL location approaches the entire chromosome, and assuming the model is correct, the QTL must lie somewhere on the chromosome.

As noted previously by Mangin *et al.* (1994), the support interval or "LOD drop-off" method of Lander and Botstein (1989) consistently underestimated the CI. Similar to the results of Darvasi *et al.* (1993), decreasing the marker spacing from 20 to 10 cM had virtually no effect on the estimated CI. The bootstrap method was also able to derive accurate CI for the other QTL parameters, such as QTL effect, and these were shown by Darvasi *et al.* (1993) to be "well behaved." Weller *et al.* (2014a) found that the nonparametric bootstrap could yield a bimodal distribution if more than a single QTL was segregating on the chromosome.

Summary

Methods to estimate parameters of QTL effects have been derived for most experimental designs of interest, and these methods are able to derive virtually unbiased estimates, based on assumptions that do not necessarily reflect reality. For example, for daughter and granddaughter designs, the general assumption is that a single QTL is segregating within a chromosomal segment, but each individual has two different QTL alleles. This will not be the case if two linked QTL are segregating in the population. Although there are analytical methods to estimate CI of QTL parameters, these methods tend to be biased, especially for QTL location. Empirical methods have been developed, which apparently work better.

12 Linkage Disequilibrium Mapping of QTL

Introduction

Even for relatively large QTL effects and sample sizes, the minimal confidence interval for QTL location will still be quite large if only linkage mapping is applied. Confidence intervals can be dramatically reduced by application of linkage disequilibrium (LD) mapping, which is based on the premise that population-wide LD exists in commercial animal populations. It is generally assumed that population-wide LD for a QTL is due to the fact that each specific QTL polymorphism was introduced only once into the population by either mutation or migration. Thus the new allele was originally associated with a specific chromosome. The fact that a segregating QTL can be detected indicates that the frequency of the rare allele increased over time either due to selection or drift. With an increase in the number of generations since the occurrence of the mutation, the length of the haplotype still associated with the mutation decreases and will differ among individuals. Unlike linkage mapping that assumes a known series of crosses, simple equations cannot be derived to estimate QTL parameters and confidence intervals for LD QTL analysis. Methods to estimate LD between markers will be considered in the first section of this chapter, and studies that estimate the extent of LD in animal populations will be reviewed in the second section. In the third section we will consider basic principles of LD QTL mapping. In the fourth section we will consider joint linkage and LD mapping of QTL as applied to the granddaughter design, and in the final section we will consider multitrait and multiple QTL LD mapping.

Estimation of Linkage Disequilibrium in Animal Populations

The two main statistics used to measure linkage disequilibrium (LD) are D' and r^2. In both cases LD is measured between each pair of loci, which we will denote A and B. D' is computed as follows:

$$D' = \sum_{i=1}^{u}\sum_{j=1}^{v} p_i q_j \left|D'_{ij}\right| \tag{12.1}$$

Genomic Selection in Animals, First Edition. Joel Ira Weller.

where u and v are the respective number of alleles at the two marker loci, p_i and q_j are the population frequencies of marker allele i at locus A and marker allele j at locus B, and $|D'_{ij}|$ = the absolute value of D'_{ij}, with $D'_{ij} = D_{ij}/D_{max}$. $D_{ij} = x_{ij} - p_i q_j$, where x_{ij} = the observed frequency of gametes $A_i B_i$, and

$$D_{max} = \begin{bmatrix} \text{Min}\left[p_i q_i, (1-p_i)(1-q_j)\right]; & D_{ij} < 0 \\ \text{Min}\left[p_i(1-q_i)(1-p_j)q_i\right]; & D_{ij} > 0 \end{bmatrix} \quad (12.2)$$

The r^2, the squared correlation of the alleles at two loci, is the preferred measure of LD for biallelic markers. In this case we will denote the two alleles at the first locus as A and a and the two alleles at the second locus as B and b. Then r^2 is computed as follows:

$$r^2 = \frac{D^2}{f(A)f(a)f(B)f(b)} \quad (12.3)$$

where $D = f(AB) - f(A)f(B)$ and $f(AB)$, $f(A)$, $f(a)$, $f(b)$, and $f(B)$ are the observed frequencies of haplotype AB and of alleles A, a, B, and b, respectively. Neither measure of LD is completely independent of allelic frequencies.

Several studies have found that population-wide LD exists in commercial animal populations, although there are conflicting reports as to its extent. This is in part due to the density and nature of markers analyzed and which statistic was used to estimate LD and the population of interest, as LD is a population-specific measurement.

Farnir *et al.* (2000) analyzing 284 microsatellites and using the D' measure of LD found that population-wide LD in dairy cattle extended in some chromosomal regions for more than 10 cM, while Sargolzaei *et al.* (2008) concluded based on analysis of 5564 SNP markers that "useful" LD ($r^2 > 0.3$) generally did not extend beyond 100 kb, or approximately 0.1 cM.

Both methods assume that the haplotypes between pairs of loci of all individuals are known without an error. Generally for individuals heterozygous for both loci, haplotypes can only be determined unequivocally if an appropriate pedigree of individuals is genotyped for these loci. In daughter or granddaughter designs, which consist of a relatively small number of half-sib families, haplotypes can be determined for the patriarch of each family based on their progeny. The paternal haplotypes of the progeny can then be determined under the assumption of zero recombination, which is reasonable if the two loci are tightly linked. There is a very extensive literature for the determination of haplotypes for more complex pedigrees (e.g., Baruch *et al.*, 2006; Weng *et al.*, 2013).

Linkage Disequilibrium QTL Mapping: Basic Principles

LD between a single marker and a QTL can be detected by regression for a biallelic marker or ANOVA for a multiallelic marker. For a biallelic marker the simplest model will be the regression of the phenotype for the quantitative trait on the number of "+" alleles (0, 1, or 2) with one of the two alleles arbitrarily determined to be the "+" allele. This method was first used successfully for a quantitative trait in animals by Cohen *et al.* (2002). Since then this method has been applied in numerous cases in many species.

Meuwissen and Goddard (2000) extended LD mapping of QTL to multiple marker loci based on haplotype analysis. Analysis of haplotypes will generally have greater statistical power than analysis of individual markers and also allows for mapping of the QTL within the haplotype. The basic assumption of the method is that for at least one of the QTL alleles, the chromosomal region in proximity to the QTL will be identical by descent (IBD) for most individuals that received this QTL allele. Thus the phenotypic values for the quantitative trait of individuals with the same haplotype in the vicinity of the QTL will be more highly correlated than individuals with different haplotypes. The basic analysis model can then be described as follows:

$$\mathbf{y} = \mathbf{Xb} + \mathbf{Zh} + \mathbf{e} \tag{12.4}$$

where **y** is the vector of records, **b** is the vector of fixed effects for which the data are to be corrected, **h** is a vector of random effects of the haplotypes, **e** is the vector of residuals, and **X** and **Z** are known incidence matrices for the effects in **b** and **h**, respectively. The variance of the residuals is Var(**e**)=$\sigma_e^2\mathbf{R}$, where **R** is assumed to be an identity matrix. The variance of the haplotype effects is Var(**h**)=$\sigma_h^2\mathbf{H}_p$, where the matrix $\mathbf{H}_p$ yields the (co)variances of the haplotype effects up to proportionality and subscript "p" indicates that $\mathbf{H}_p$ depends on the assumed position of the QTL. The dimension of $\mathbf{H}_p$ is $q*q$, where q is the number of different haplotypes in the data set given the haplotype length.

Assuming multivariate normality, the residual log-likelihood of the data under the above model is

$$\mathbf{L}\left(\mathbf{H}_p,\sigma_h^2,\sigma_e^2\right) \propto -0.5\left[\ln\left|\mathbf{V}\right| + \ln\left|\mathbf{X}'\mathbf{V}^{-1}\mathbf{X}\right| + \left(\mathbf{y}-\mathbf{X}\hat{\mathbf{b}}\right)'\mathbf{V}^{-1}\left(\mathbf{y}-\mathbf{X}\hat{\mathbf{b}}\right)\right] \tag{12.5}$$

where $\mathbf{V}$=Var(**y**)=[$\mathbf{ZH}_p\mathbf{Z}'\sigma_h^2+\mathbf{R}\sigma_e^2$] and $\hat{\mathbf{b}}$ is the generalized least-squares estimate of **b**. Given a QTL position, p, that is, given $\mathbf{H}_p$, this likelihood is maximized to obtain estimates of the variance components σ_h^2 and σ_e^2. The elements of $\mathbf{H}_p$ are the probabilities that the two haplotypes corresponding to each row and column of the matrix received the same QTL allele IBD times σ_h^2. That is, the covariance between two haplotype effects, h_i and h_j, is

$$\mathrm{Cov}\left(h_i, h_j\right) = \mathrm{Prob}\left(\mathrm{IBD} \mid \text{marker haplotypes}\right) \times \sigma_h^2 \tag{12.6}$$

where Prob(IBD|marker haplotypes) is the probability that the QTL locus is IBD given the marker haplotypes. Calculation of these probabilities is not trivial, but can be computed either by the "coalescence process" (Hudson, 1985) or the "gene dropping" method (Maccluer *et al.*, 1986). Both methods require extensive computations.

In the gene dropping method, markers and QTL are simulated in a base generation of N_e individuals. All $2N_e$ base QTL alleles, which are called founder alleles, have a unique number. The following N_G descendant generations are simulated by choosing at random parents from the previous generation and letting their N_e offspring inherit haplotypes or recombinant haplotypes according to Mendel's rules and the recombination probabilities. Because all the founder QTL alleles have unique numbers, any two QTL alleles with the same number in generation N_G are IBD.

The IBD probabilities of a pair of haplotypes can be estimated within each simulation by dividing the number of times the QTL locus was IBD by the total number of times the haplotype pair was found. The estimates of the IBD probabilities of the haplotype pairs that belong to the same haplotype pair group are averaged within a simulation run, and these averages are accumulated across 100,000 repeated simulations.

Applying this method to the simulated data, a QTL was correctly positioned within a region of 3, 1.5, and 0.75 cM in 70, 62, and 68%, respectively, of the replicates using markers spaced at intervals of 1, 0.5, and 0.25 cM, respectively.

Joint Linkage and Linkage Disequilibrium Mapping

In a daughter or granddaughter design, paternal haplotypes of the final generation genotyped are identical to the haplotypes of their sires, except for recombination, while the maternal haplotypes can be considered a random sample from the population. For linkage mapping of daughter or granddaughter designs described previously, only the paternal haplotypes are analyzed. In order to utilize both haplotypes, Meuwissen *et al.* (2002) presented an algorithm for joint linkage and LD mapping. The first step is determination of haplotypes, which should generally not be a problem for daughter or granddaughter designs. Meuwissen *et al.* (2002) used a Gibbs sampling algorithm but included only those haplotypes that could be determined with near certainty.

Similar to the case in the previous section, IBD probabilities at the assumed QTL location must be computed between all pairs of haplotypes. The IBD probabilities at the QTL of the base haplotypes with the paternal haplotypes of the sons, and among the paternal haplotypes, are obtained from the following equation, which states that the IBD probability, $P_{IBD}(X(p);Y)$, of the paternal QTL allele of son X, $X(p)$, with any other QTL allele, Y, equals

$$P_{IBD}\left(X(p);Y\right)=r\left[P_{IBD}\left(S(p);Y\right)\right]+(1-r)\left[P_{IBD}\left(S(m);Y\right)\right] \tag{12.7}$$

(Fernando and Grossman, 1989), where $S(p)$ and $S(m)$ denote the paternal and maternal alleles of the sire S, respectively, and r is the probability that the son inherited the paternal QTL allele of the sire. Hence, $X(p)=S(p)$ with probability r, and $X(p)=S(m)$ with probability $(1-r)$. The probability r was predicted from the paternal or maternal inheritance of the nearest informative markers that flanked the putative QTL position. The above equation is used recurrently to fill in the missing IBD probabilities at the QTL of paternal haplotypes of sires using the known IBD probabilities among the base haplotypes.

The next step is then to model the records as a function of the haplotype effects. In the case of the granddaughter design, the "records" analyzed will be the sons genetic evaluations or DYD, as explained previously. Since the sons are related through sires, Equation (12.4) was modified as follows:

$$\mathbf{y}=\mu\mathbf{1}+\mathbf{Zh}+\mathbf{u}+\mathbf{e} \tag{12.8}$$

where μ is the overall mean, **1** is a vector of ones, **u** is a vector of random polygenic effects, and the other terms are as in Equation (12.4). The variance of the haplotype effects is as described, and the variance of the polygenic effects is $\mathbf{A}\sigma_s^2$, where **A** is the numerator relationship matrix and σ_s^2 is the variance among the genetic evaluations or DYD. This model differs from the model of Equation (12.4) in that fixed effects, other than a population mean, are not included, while a polygenic effect is. As in the previous section, the most likely QTL location is determined by maximizing the residual log-likelihood relative to the QTL position. The residual log-likelihood will also be slightly different, because of the inclusion of a polygenic effect and the deletion of fixed effects other than a general mean.

Using this method, Meuwissen *et al.* (2002) were able to map a QTL affecting twinning rate to a chromosomal region of less than 1 cM in the middle part of bovine chromosome 5. Olsen *et al.* (2005) used this method to map the large QTL on bovine chromosome 6 affecting protein concentration to a region of 420 kb, approximately 0.5 cM. With linkage mapping via a granddaughter design, the confidence interval for QTL location was 7.5 cM. A few months later, Cohen-Zinder *et al.* (2005) identified the causative polymorphism for this QTL as a missense mutation in the ABCG2 gene located within the region proposed by Olsen *et al.* (2005).

Multitrait and Multiple QTL LD Mapping

The method of joint linkage and LD mapping was extended by Meuwissen and Goddard (2004) to multitrait and multiple QTL analysis. Assuming that m traits are analyzed, the vector of m phenotypic records of animal i, $\mathbf{y}_i$, is modeled by

$$\mathbf{y}_i = \mathbf{X}_i\mathbf{b} + \mathbf{u}_i + \sum_j \left(q_{ji1} + q_{ij2}\right)\mathbf{v}_j + \mathbf{e}_i \tag{12.9}$$

where $\mathbf{y}_i$ here is the $(m'1)$ vector of daughter yield deviations (DYD) of sire i, $\mathbf{X}_i\mathbf{b}$ denotes the $(m'1)$ vector of (nongenetic) fixed effect corrections for the traits of animal i, $\mathbf{u}_i = (m'1)$ vector of effects of the background genes (polygenic effect) on each of the traits, $\mathbf{e}_i = (m'1)$ vector of environmental effects on each of the traits, Σ_j denotes summation over all possible QTL positions on the chromosome, $\mathbf{v}_j$ = the $(m'1)$ direction vector of the direction of the effects of the QTL alleles on different traits at position j, and q_{ij1} (q_{ij2}) = the size of the QTL effect for the paternal (maternal) allele of animal i at position j along the direction $\mathbf{v}_j$.

The dependencies between the effects of the fitted QTL are reduced by assuming that there is only one QTL per marker bracket and that only the midpoints of the brackets are considered as putative QTL positions. The likelihood conditional on all unknowns was assumed to be multivariate normal. Equations for the complete joint posterior distribution are rather complicated and are given in Meuwissen and Goddard (2004). Parameters were estimated by Gibbs sampling. The method was applied to the QTL on bovine chromosome 14 also analyzed by Riquet *et al.* (1999). The quantitative trait nucleotide for this QTL was identified as a missense mutation in the gene DGAT1 (Grisart *et al.*, 2002; Winter *et al.*, 2002), which apparently affects all milk production traits but has the greatest effect on fat concentration. The QTL was mapped to a region of 0.04 cM, and the effects of the gene were accurately estimated as compared to previous studies. No indications for a second QTL affecting milk production traits were found on this chromosome, despite the results of Kuhn *et al.* (2004) who found that additional polymorphisms in DGAT1 also affect fat concentration.

Summary

Similar to linkage mapping, LD mapping was first developed to map specific QTL and estimate their effects. Although with application of LD mapping it is possible to map QTL to chromosomal intervals of less than a single map unit, its current relevance to genomic selection is very limited, as will be demonstrated in the coming chapters. Even though genomic selection is based on LD between genetic markers and QTL, determination of the effects and location of the specific genes underlying the economic traits are generally not considered as requirements for computation of genomic estimated breeding values.

13 Marker-Assisted Selection: Basic Strategies

Introduction

In this chapter we summarize the literature dealing with marker-assisted selection (MAS) prior to the introduction of high-density marker panels including tens of thousands of markers. Most of the literature dealing with MAS prior to 2000 assumed that a relatively low number of chromosomal segments were assumed to contain QTL of interest, although some of the theory is independent on the number of markers included in the analysis.

We will first review the situations in which selection index is inefficient, and in the next section we will present the general considerations for MAS within a breed. In the following section we will consider the specific problems of MAS in segregating populations. Formulas to compute the optimum selection index with phenotypic and marker information and to compare phenotypic selection and MAS for individual selection will be presented in the following section. MAS with traits expressed only in a single sex, and selection on juveniles will be considered in the next two sections. Optimization of MAS with family selection will be considered in the following section, and the reduction of selection gain with MAS due to sampling will be considered in the next section. Problems of MAS related to segregating problems will be considered in the next two sections. In the final sections we will consider genetic evaluations based on dense whole genome scans, and predicted genetic gains obtained in simulation studies.

Situations in Which Selection Index is Inefficient

The practical situations in which selection index is not efficient can be listed as follows:

1. Low heritability for trait included in the economic objective
2. Traits that cannot be scored on all individuals (males, juveniles, live animals, disease challenge)
3. Negative genetic correlations among traits
4. Nonadditive genetic variance (dominance, epistasis)
5. Epigenetic effects
6. Crossbreeding
7. "Cryptic" genetic variation
8. Introgression

Genomic Selection in Animals, First Edition. Joel Ira Weller.

Many traits of major economic importance have been neglected in breeding programs because of low heritability. Prime examples are fertility traits and disease resistance. Selection index works best on traits with near normal continuous distributions. Thus, traits such as conception rate, number of progeny, or disease have received less emphasis in breeding programs. Selection index is less efficient when the trait is expressed only in one sex or only in mature individuals. Certain traits cannot be scored on live animals, such as carcass composition. In this case genetic values can only be estimated through records of relatives.

As shown by Falconer (1964) negative genetic correlations among traits included in the selection objective tend to build up over time. Nearly all commercial breeding programs include traits with negative genetic correlations. The effect of negative genetic correlations among traits included in the selection objective will be considered later in detail.

Clearly, selection index does not utilize nonadditive genetic variance, nor does it utilize epigenetic variance. "Epigenetics" refers to functionally relevant changes to the genome that do not involve a change in the nucleotide sequence. Examples of mechanisms that produce such changes are DNA methylation and histone modification, each of which alters how genes are expressed without altering the DNA base pair sequence of the individual. Epigenetic changes have been observed to occur in response to environmental effects and may "mutate" several times per generation or may be fixed for many generations.

Selection index theory does not provide an answer for crossbreeding among traits. The three main goals of crossbreeding are (i) utilization of heterosis, (ii) increased genetic variation, and (iii) introgression. The "classical" explanations for heterosis are elimination of inbreeding depression and overdominance at the level of the individual locus. Even in the absence of these "true" genetic effects, crossbreeding is often more profitable than selection within a single line. Moav (1966) defined five types of "economic" heterosis.

Different breeds are sometimes crossed to produce a population with increased genetic variance. Selection index can then be used to increase the economic value in future generations. However, desirable genes of individuals with overall inferior phenotypes can be lost through trait-based selection. Generally only the economically best breeds will be considered as parental candidates. Again, some breeds with overall inferior phenotypes may carry some desirable genes, which will not be found by trait-based selection. This is especially true of wild progenitors of domestic species. This "cryptic" genetic variation can be utilized via MAS.

Potential Contribution of MAS for Selection within a Breed: General Considerations

Potentially, MAS can increase annual genetic gain by:

1. Increasing the accuracy of evaluation
2. Increasing the selection intensity
3. Decreasing the generation interval

Most of the studies on MAS have dealt with increasing the accuracy of evaluation. Information on the individual genes affecting the trait of interest does increase the accuracy of the evaluation, but the effect decreases as the heritability increases. Assume that marker information is available for QTL affecting some of the traits included in the breeding objective. We will define m_s as the "net marker score," which is the sum of the additive effects associated with the markers for a given individual. With information on individual loci in addition to phenotypic trait values, Lande and Thompson (1990) proposed that selection index methodology can be used to construct an optimum linear selection index, I_s, of the form

$$I_s = \mathbf{b}'_y \mathbf{y} + b_m m_s \tag{13.1}$$

where $\mathbf{b}_y$ represents the index coefficients for the quantitative trait records, $\mathbf{y}$, and b_m represents the index coefficient for m_s. $\mathbf{b}_y$ and $\mathbf{y}$ are vectors, while b_m and m_s are scalars. That is, the marker information can be considered the addition of a single trait to the selection index. The index coefficients can be computed based on the general selection index equations:

$$\mathbf{b} = \mathbf{V}_p^{-1}\mathbf{C}\mathbf{v} \tag{13.2}$$

where $\mathbf{V}_p$ is the phenotypic variance matrix for the recorded traits, $\mathbf{C}$ is the genetic covariance matrix between the recorded traits and the traits with economic values, and $\mathbf{v}$ is the vector of economic values. In the case of selection on phenotype and marker information, the marker score has no intrinsic economic value. Therefore, the coefficient of the net marker score in $\mathbf{v}$, the vector of economic values, will be equal to zero. We will now consider in detail several situations of interest.

Phenotypic Selection versus MAS for Individual Selection

In the simplest case we will assume that for trait-based selection individuals are selected based on a single phenotypic record and that for MAS individuals are selected based on the phenotypic record and their own marker information. Information from relatives is not considered. The phenotypic and genetic variance matrices are computed as follows:

$$\mathbf{V}_p = \begin{bmatrix} \sigma_p^2 & \sigma_m^2 \\ \sigma_m^2 & \sigma_m^2 \end{bmatrix} \quad \text{and} \quad \mathbf{G} = \begin{bmatrix} \sigma_A^2 & \sigma_m^2 \\ \sigma_m^2 & \sigma_m^2 \end{bmatrix} \tag{13.3}$$

where σ_p^2 and σ_A^2 are phenotypic and genetic variances and σ_m^2 is the additive genetic variance explained by the genetic markers. In terms of the additive genetic variance, these equations become

$$\mathbf{V}_p = \begin{bmatrix} 1/h^2 & p_m \\ p_m p_m & \end{bmatrix} \quad \text{and} \quad \mathbf{G} = \begin{bmatrix} 1 & p_m \\ p_m p_m & \end{bmatrix} \tag{13.4}$$

where h^2 is the heritability and p_m is the fraction of the additive genetic variance associated with the genetic markers, that is, $p_m = \sigma_m^2/\sigma_A^2$. Inverting $\mathbf{V}_p$ and substituting into Equation (13.2) gives index coefficients of $(p_m - p_m^2)(p_m/h^2 - p_m^2)$ and $(p_m/h^2 - p_m^2)^2$. The actual b-values will be functions of the trait units. Therefore the ratio of the values has more intrinsic meaning. This ratio is computed as follows:

$$\frac{b_m}{b_x} = \frac{1/h^2 - 1}{1 - p_m} \tag{13.5}$$

where b_m and b_x are the index coefficients for marker and phenotypic information, respectively. From this equation it can be deduced that as the heritability of the selection objective tends toward unity, b_m tends to zero, regardless of p_m.

The relative selection efficiency (RSE) of two different indices is defined as the ratio of their expected genetic gains (Weller, 1994). The economic value of genetic gain by phenotypic selection will be $i_p h\sigma_A$. Thus the RSE of a selection index including marker information to a selection index based only on trait values for individual selection will be equal to $(\mathbf{v}'\mathbf{G}'\mathbf{V}_p^{-1}\mathbf{G}\mathbf{v})^{0.5}/(h\sigma_A)$. The elements of **v**, **G**, and $\mathbf{V}_p$ are given earlier. $\mathbf{V}_p$, a 2×2 matrix, can be easily inverted. Inverting this matrix and multiplying the vectors and matrices gives (Lande and Thompson, 1990)

$$\text{RSE} = \left[\frac{p_m}{h^2} + \frac{(1-p_m)^2}{1-h^2 p_m}\right]^{0.5} \tag{13.6}$$

As heritability tends to unity, so does RSE. For $h^2=0.25$ and $p_m=0.5$, RSE = 1.5. Thus, gains for individual selection through MAS can be quite significant. Equation (13.6) gives the added gain due to selection on an index including phenotypic information on the selection objective and marker information. If selection is based only on known QTL without information on the economic trait, then the RSE of MAS to trait-based selection is $(p_m/h^2)^{½}$. Thus, selection efficiency on markers alone will be greater than trait-based selection if $p_m > h^2$.

MAS for Sex-Limited Traits

As noted previously, selection index is inefficient in situations in which the selection criteria cannot be scored on all individuals, for example, a sex-linked trait. Selection efficiency can be increased by selection among individuals without phenotypic expression of trait in addition to among those with phenotypes. For example, milk production is expressed only in females. Therefore selection among males is based only on information from relatives. With only information on relatives, two full brothers will have the same genetic evaluation. Information on markers could be used to differentiate between them. Furthermore, in many animal species, although the traits of interest are expressed only in females, females have a low fertility rate, while males have a very high potential fertility rate. Thus, the selection intensities will also be different in the two sexes. For a trait expressed only in females, the RSE of MAS on both sexes relative to individual phenotypic selection of females will be

$$\text{RSE} = \left[\frac{p_m}{h^2} + \frac{(1-p_m)^2}{1-h^2 p_m}\right]^{0.5} + \frac{i_m}{i_f}\left[\frac{p_m}{h^2}\right]^{0.5} \tag{13.7}$$

where i_m is the selection intensity in males and i_f is the selection intensity in females. The first term of this equation is the same as the previous equation and refers to selection of females for which both marker and phenotypic information are available. The second term refers to selection on males, for which only marker information can be used.

In this case RSE can be significantly higher, as compared to situations in which the trait is expressed in both sexes. For example, if $p_m = h^2$ and $i_m/i_f = 2$, then RSE is doubled, relative to the situation in the previous equation. As heritability tends toward unity, this equation tends to $1+\left(i_{\mathrm{f}}/i_{\mathrm{m}}\right)\sqrt{p_{\mathrm{m}}}$. The maximum RSE as p_{m} tends toward unity for any heritability is $(1+i_{\mathrm{f}}/i_{\mathrm{m}})/h$.

MAS Including Marker and Phenotypic Information on Relatives

With both marker and trait information on both the individual and his relatives, selection index theory can again be used to construct the optimum selection index, which will have the following form:

$$I_{\mathrm{s}}=b_{\mathrm{zf}}Z_{\mathrm{f}}+b_{\mathrm{mf}}m_{\mathrm{f}}+b_{\mathrm{zw}}Z_{\mathrm{w}}+b_{\mathrm{mw}}m_{\mathrm{w}} \tag{13.8}$$

where Z_{f} is the mean family phenotype, m_{f} is the mean family marker score, Z_{w} is the phenotypic deviation of the individual from the family mean, m_{w} is the deviation of the individuals molecular score from the family mean, and the b's are the appropriate index coefficients. As in individual selection it is assumed that the marker scores have no intrinsic economic values. We will further assume that the selection objective is measured in units of its economic value. In this case, the vector of economic weights will be [1 0 1 0]′.

We will define r_{f} as the fraction of genes identical by descent among family members (½ for full-sibs, and ¼ for half-sibs), n as the number of individuals in each family, and c^2 as the residual correlation among family members. The values for the index coefficients can be derived based on selection index theory as follows (Lande and Thompson, 1990):

$$\begin{bmatrix} b_{\mathrm{zf}} \\ b_{\mathrm{mf}} \\ b_{\mathrm{zw}} \\ b_{\mathrm{mw}} \end{bmatrix}=\begin{bmatrix} r_{\mathrm{n}}h^2\left(1-p\right)/D_{\mathrm{f}} \\ \left(t_{\mathrm{n}}-r_{\mathrm{n}}h^2\right)/D_{\mathrm{f}} \\ \left(1-r_{\mathrm{f}}\right)h^2\left(1-p\right)/D_{\mathrm{w}} \\ \left[1-t-\left(1-r_{\mathrm{f}}\right)h^2\right]/D_{\mathrm{w}} \end{bmatrix} \tag{13.9}$$

where $r_{\mathrm{n}}=r_{\mathrm{f}}+(1-r_{\mathrm{f}})/n$, $t=r_{\mathrm{f}}h^2+c^2$, $t_{\mathrm{n}}=t+(1-t)/n$, $D_{\mathrm{f}}=t_{\mathrm{n}}-r_{\mathrm{n}}h^2p$, and $D_{\mathrm{w}}=1-t-(1-r_{\mathrm{n}}h^2p)$. The expression for RSE using information on relatives is quite complex and is given in Lande and Thompson (1990).

Maximum Selection Efficiency of MAS with All QTL Known, Relative to Trait-Based Selection, and the Reduction in RSE Due to Sampling Variance

The maximum RSE that can be obtained for various selection schemes with $p=1$ was also computed by Lande and Thompson (1990). Very large families are assumed for the combined individual and family selection schemes. The RSE computed for selection based on half-sib or full-sib records is much less than for individual phenotypic selection. With half-sib selection, the maximum gain possible, as p tends toward unity, is $2[(1-h^2/4)/(1+2h^2)]^{1/2}$. For $h^2=0.5$, maximum RSE = 1.32, as compared to a RSE of 2 for individual selection with the same heritability.

In all of the previous equations, RSE was estimated under the assumption that QTL effects were estimated without error. However, if the sample size is finite, there will be sampling errors in the estimated QTL effects. The loss in selection efficiency for MAS due to sampling error will be approximately equal to the following expression (Lande and Thompson, 1990):

$$\frac{\left(2h^2 p_{\mathrm{m}} + N_{\mathrm{Q}}/N\right) p_{\mathrm{m}} \left(1-h^2\right)^2}{Nh^2\left(1-p_{\mathrm{m}}h^2\right)\left[p_{\mathrm{m}} + h^2\left(1-2p_{\mathrm{m}}\right)\right]^2} \tag{13.10}$$

where N is the number of individuals analyzed and N_{Q} is the number of marker loci included in the selection index. The reduction in RSE will be less than 2% if at least a few hundred individuals are analyzed, for any combination of p_{m} and h^2 (Lande and Thompson, 1990).

Marker Information in Segregating Populations

Even if segregating QTL are detected via linkage to genetic markers, there are two major problems that must be addressed if this information is to be included in actual breeding programs:

1. Linkage phase can be different in different individuals. Thus it will be necessary to determine the QTL alleles and phase for each candidate for selection.
2. Unless the markers are very tightly linked to the QTL, linkage relationships will break down in future generations.

To overcome both of these problems, a number of studies have assumed that the actual QTL have been identified, and the effects of the different alleles are known *a priori*. Once the QTL effect is determined, it is necessary only to genotype candidates for selection to determine their QTL genotypes. So far only four QTN have been identified in commercial animal populations (Ron and Weller, 2007). More recent studies also indicate that the number of QTN that can be identified in dairy cattle is probably less than 10 (https://www.cdcb.us/Report_Data/Marker_Effects/marker_effects.cfm?Breed=HO). Considering these limitations justifications for QTN determination were summarized by Weller and Ron (2011).

Inclusion of Marker Information in "Animal Model" Genetic Evaluations

Most studies that have evaluated MAS have generally assumed that the genome is first scanned to locate chromosomal regions containing QTL. Using additional markers, the QTL are progressively localized to smaller and smaller chromosomal regions, and finally the actual genes are identified. The identified QTL are then used in selection programs (Soller, 1994). Following this approach, or even localization of the QTL to a very small chromosomal segment, recombination in future generations is no longer a problem, but there is a significant time lag until QTL are utilized in breeding programs.

An alternative approach was presented by Fernando and Grossman (1989). Their model estimates breeding values of all individuals in the population, including information from genetic markers, but does not directly estimate the QTL effects. Instead, they modified a standard individual animal model so that in addition to the polygenic effect of each individual, two "gametic effects" are estimated for the two parental marker alleles or haplotypes passed to each individual for each locus.

Rather than representing specific QTL alleles, these gametic effects include uncertainty with respect to the QTL allele received. Following the principles of selection index, selection based on the estimated breeding values including marker information should result in maximum genetic gain in the next generation, even though QTL information is incomplete.

Israel and Weller (1998) proposed a complete mixed model analysis of the population with a fixed genotype effect for all individuals, including individuals that were not genotyped. For these individuals the coefficients of the genotype effect are the probability of each possible QTL genotype, based on allele frequencies in the population, and known genotypes of relatives. However, when this model was applied to actual data from the Israeli Holstein population for the *DGAT1* locus segregating QTL on chromosome 14 that affected milk production traits (Grisart *et al.*, 2002), the QTL effect was strongly underestimated relative to alternative estimation methods. This bias is apparently due to the fact that the genotype probabilities tend to "mimic" the effect of relationships as the fraction of animals with inferred genotypes increases. Baruch and Weller (2008) were able to derive unbiased estimates of quantitative trait locus effects by the following modified "cow model":

$$Y_{ijk} = c_i + h_j + m_k + q + e_{ijk} \tag{13.11}$$

where c_i = random effect of cow i, h_j = the effect of herd-year-season j, m_k = the fixed parity effect, q = the QTL substitution effect, and e_{ijk} = the random residual effect. This model differs from the model of Israel and Weller (1998) in that only cows with production records are included, and covariances among cow effects are assumed to be zero. That is, the relationship matrix is not included.

Although this method can be used to derive unbiased QTL estimates, it cannot be used to derive genetic evaluations, as animals without records, including all males, are not included.

Predicted Genetic Gains with Genomic Estimated Breeding Values: Results of Simulation Studies

In this section we present the results of simulation studies that assume a given accuracy for genomic estimated breeding values (GEBV) and further assume that GEBV are unbiased relative to the true genetic values. Simulation studies that evaluate GEBV will be presented in Chapter 17.

Most studies have compared genomic selection schemes with the conventional progeny test scheme diagrammed in Figure 3.2 (Bouquet and Juga, 2013). Two basic strategies have been proposed for incorporation of genomic selection into commercial breeding programs:

1. The "preselection" scheme entails using genomic information to preselect young males for a progeny test. All subsequent steps for the selection of males remain the same. This scheme increases the rate of genetic gain by increasing the selection accuracy of young male candidates.
2. In the "juvenile" schemes, AI sires for general service are selected based on GEBV among young genotyped males able to produce semen. Although the GEBV of young sires are less accurate than conventional breeding values estimated for progeny-tested bulls, the loss in selection accuracy is compensated by a huge reduction in generation intervals, as the progeny testing step is eliminated. Thus this scheme is similar to the "half-sib" selection scheme diagrammed in Figure 3.1, except that bulls are selected based on GEBV rather than only on pedigree.

Large variations in predicted annual genetic gain were found across studies, as described by Pryce and Daetwyler (2012). Compared with conventional schemes, gains in annual genetic gain ranged from +9% for preselection schemes to more than +100% for juvenile schemes. Increases in genetic gain were largest for selection on low heritability traits, because genomic data added relatively more information to predict breeding values for these traits. Lillehammer *et al.* (2011) found that annual genetic gain was increased by 29, 40, and 70% in juvenile schemes for heritability values 0.30, 0.05, and 0.01, respectively.

With genomic evaluation, the accuracy of genotyped females should approach the accuracies obtained for genotyped males. In addition, GEBV of bull dams are expected to be less biased, because genomic information reduces the weight attributed to the cows' own production records, which may be subject to preferential treatment. With a training population of sufficient size, large gains are expected from a more accurate selection of breeding cows, even when only a small proportion is genotyped (Sørensen and Sørensen, 2009). They showed that allocating larger proportions of genotypings to females as opposed to males resulted in larger selection responses.

Finally, genomic selection schemes should reduce rates of increase in inbreeding. With GEBV the weight of family information in genetic evaluation is reduced by placing emphasis on Mendelian sampling information (Daetwyler *et al.*, 2007; Dekkers, 2007). For example, two full brothers without progeny records will have equal genetic evaluations based only on pedigree but different GEBV based on their individual Mendelian samplings of their parents' genotypes. The largest reductions in inbreeding rates due to the use of genomic selection were observed for traits of low heritability (Lillehammer *et al.*, 2011) and when a large part of variance was explained by markers (deRoos *et al.*, 2011). On the other hand, Buch *et al.* (2012) found that the annual rate of increase in inbreeding was slightly greater for juvenile scheme as compared to the standard progeny test but lower for the preselection scheme. This is due to the fact that the mean generation interval is lower for the juvenile scheme. Per generation both genomic selection schemes had lower inbreeding than the progeny test scheme.

Summary

Although trait-based selection is very efficient in certain situations, in many practical cases, this is not the case, and these situations are summarized in this chapter. Formulas were presented that can be used to evaluate the relative efficiency of MAS, as compared to traditional trait-based selection, for a number of situations of interest. In some cases the selection efficiency of MAS can potentially be more than 1.5 times traditional selection. As the situations considered approach reality, the formulas become more complicated, and more parameters must be considered. Most situations of real-world interest cannot be evaluated analytically, and simulation of these scenarios is required. It should be noted though that even with the advent of high-density marker panels the question of how to correctly weight marker information with pedigree and phenotypic information is still one of the central problems for practical application of MAS.

14 Genetic Evaluation Based on Dense Marker Maps: Basic Strategies

Introduction

Current "mid-density" SNP chips generally include over 50,000 genetic markers. A "high-density" chip with more than 777,000 markers is also available (e.g., http://www.illumina.com/products/bovinehd_whole-genome_genotyping_kits.ilmn). As the number of markers increases into the tens of thousands, there will be population-wide linkage disequilibrium (LD) between markers and closely linked QTL. Thus it should be possible to detect nearly all of the segregating genes with effects on the traits of interest. Based on this assumption, genetic evaluations based on this number of markers are termed "genomic evaluations."

In this chapter we will consider the basic question related to genomic evaluation. For genomic evaluations the multiple comparison problem considered previously becomes more acute. A second problem is that most markers will have no measurable effect on the traits analyzed. There the "effects" estimated for most markers will be merely "noise." Thus various strategies have been proposed to include only markers with actual effects in genomic evaluation algorithms.

We will also consider the question of whether marker effects should be considered fixed or random, and whether individual markers or haplotypes should be analyzed. We will then consider the criteria used to determine which markers should be included in the analyses. Finally we will describe how a genomic variance matrix can be constructed and the proposed methods for evaluation of genomic estimated breeding values (GEBV).

The Basic Steps in Genomic Evaluation

Goddard and Hayes (2007) proposed that genomic selection should be considered a three-step process:

1. Use the markers to deduce the genotype of each animal at each QTL.
2. Estimate the effects of each QTL genotype on the trait.
3. Sum all the QTL effects for selection candidates to obtain their genomic EBV (GEBV).

Genomic Selection in Animals, First Edition. Joel Ira Weller.

As explained in the previous chapter, nearly all recent studies have also proposed a fourth step: construction of a selection index that incorporates the direct genomic evaluations with information from pedigrees and phenotypic records.

Evaluation of Genomic Estimated Breeding Values

In general two methods have been proposed to evaluate the accuracy of GEBV. In the first method, first employed by Meuwissen *et al.* (2001), GEBV are computed from simulated data sets. In this case the simulated breeding values (BV) are known and can be compared to GEBV computed from the marker and phenotypic data of the simulated data sets. The advantage of this method is that the "true" genetic values are known, and any number of simulated data sets can be generated. The disadvantage is that we never know how accurately the simulation algorithm corresponds to actual data.

Goddard and Hayes (2007) proposed the following procedure to evaluate genomic selection methodologies. A prediction equation that uses markers as input and predicts BV is derived from a "discovery" data set where a large number of SNPs have been assayed on a sample of animals with phenotypes for all the relevant traits. Then the accuracy of the prediction equation is evaluated on an independent "validation" data set in which a second sample of animals are recorded for the traits and genotyped at least for the markers that are proposed to be used commercially. In the case of sires, the GEBV predicted in the validation data set based only on pedigree and marker data are compared to EBV for the same animals based on progeny test derived by standard BLUP evaluations.

Generally the basis for comparison of GEBV computed for young animals is the EBV of these animals computed based only on pedigree data. If both parents have EBV computed by an animal model, then the EBV of the young bulls will be the mean of their parent EBV. Selection candidates are genotyped for the markers, and the prediction equation estimated in the discovery data is used to calculate GEBV, but their accuracy is assumed to be that found in the validation sample. Criteria for evaluation of GEBV will be discussed in detail in Chapter 17.

Sources of Bias in Genomic Evaluation

Generally bulls with genetic evaluations based on their daughters will be genotyped. With respect to estimation of QTL effects in dense genome scans by linear models, there is one source of upward bias in the estimation of QTL effects, and two sources of downward bias:

1. If multiple QTL are estimated as fixed effects, the estimated effects of those QTL that meet the "significance" criterion will be biased upward due to selection. This has been denoted the "Beavis effect" (Beavis, 1994).
2. QTL effects estimated from genetic evaluations or daughter yield deviations (DYD) will be biased downward (Israel and Weller, 1998).
3. Unless the actual QTN has been detected, the QTL effect will be underestimated due to incomplete LD between the linked markers and the QTL.

With respect to the first problem, this is one of the main reasons that most recent studies have assumed that marker-associated effects should be considered random. The problem of bias with analysis of DYD or genetic evaluations as the dependent variable will be considered in detail in the following two chapters. With respect to the third problem, the proportion of the QTL variance explained by the markers, r^2, is dependent on the LD between the QTL and the marker, or a linear combination of markers if haplotypes are analyzed. The extent of LD and hence r^2 are highly variable across the genome, but r^2 declines as the distance between the two loci increases. In Holstein cattle average r^2 between loci 50 kb apart was estimated at 0.35 (Goddard *et al.*,2006). To obtain an average spacing of 50 kb requires 60,000 evenly spaced markers.

Marker Effects Fixed or Random?

At first glance, it would seem that effects associated with the markers could be considered as fixed. Only additive genetic variance is generally useful for within-breed genetic improvement. Therefore, nonadditive genetic variance is generally ignored, both within loci (i.e., dominance) and among loci (i.e., epistasis). Thus, most models have assumed that the marker effect is a simple linear regression on the number of "+" alleles. Since generally there are only two alleles for each SNP, only a single contrast between these two alleles has to be estimated for each allele of each SNP. Furthermore the sample size will in any event be very large. Therefore there should be virtually no "shrinkage" of the SNP effects, due to regression toward the mean, in a standard BLUP model in which each SNP is considered a separate random effect.

However, despite these considerations, nearly all studies have assumed SNP effects to be random for the following reasons:

1. If the analysis model includes a single record for each animal genotyped, then the number of markers will generally be much larger than the number of records included in the analysis. In this case if the marker effects are considered to be fixed, the model will be "overparameterized." That is, the number of effects estimated is larger than the number of records to be fitted, and infinitely many "solutions" will fit the data without error.
2. If the marker effects are fixed, it is not possible to account for covariances among the marker effects. Clearly with a dense marker map, more than one marker will be in linkage disequilibrium with a segregating QTL. This will result in covariances among markers.
3. Most markers will have no effect on any given trait. Thus the effects estimated for these markers will be merely "noise." This problem is greater if markers are considered fixed effects, in which case these "pseudo" effects are not regressed.

A few studies have nevertheless considered markers as fixed effects (Baruch and Weller, 2008, 2009). Overparameterization is not a problem if the number of records is greater than the number of markers, and this can be achieved by including all animals with records in the analysis, even though the vast majority of animals do not have genotypes. For a fixed effect model, the genotypes for individuals that were not genotyped are replaced by the probabilities of each possible genotype, based on the known genotypes of relatives and allelic frequencies in the population. The problem that most markers do not affect any specific trait can be alleviated by preselection of markers, which can also solve the overparameterization problem. Strategies for analysis of all animals in the population with markers assumed to be random and all markers included in the analysis will be considered in detail in the following chapter.

Individual Markers versus Haplotypes

The analysis models described so far computed individual effects for each marker. Since markers are not transmitted individually, but as part of chromosomal block, it would seem that analysis of haplotypes should yield more accurate genomic evaluations. Goddard (1992) proposed that it should be possible to increase the proportion of genetic variance captured at the QTL by using haplotypes of markers, rather than single markers, to identify the QTL alleles carried by each animal. However, in practice this is not the case, and no efficient method for genomic evaluation based on analysis of haplotypes has been described in the literature. It should also be noted that a genomic evaluation model based on analysis of haplotypes would be intrinsically more complicated, because for each haplotype segment there would be multiple possible haplotypes, and it is necessary to estimate an effect for each one.

Kolbehdari *et al.* (2006) developed a linear regression on haplotype model for mapping QTL in half-sib designs. They compared haplotypes consisting of either two or four markers. Empirical power with four-marker haplotypes was the same or greater than in two-marker intervals, but the probabilities of type I errors were slightly greater, and the precision was lower (bigger absolute differences from the true position) with analysis of single markers compared to two-marker intervals. Thus on a practical level it does not seem that a significant gain can be obtained by analysis of haplotypes, as opposed to individual markers.

Total Markers versus Usable Markers

So far all analyses on actual data have not used all markers included on SNP chips. The main reasons why markers are deleted are as follows:

1. Low call rate. In analysis of actual data, it is generally found that genotypes cannot be reliably determined for most individuals in between 5 and 10% of the markers included in commercial BeadChips. Nearly all studies have set a minimum "call rate" for inclusion of markers in the analysis.
2. Lack of polymorphism. In any specific population analyzed a certain fraction of markers will all share the same homozygous genotype, or frequency of the second allele will be so low as to render the marker virtually useless. Most studies have required a 5% threshold for the minor allele frequency (e.g., VanRaden *et al.*, 2009).
3. Marker redundancy. If markers are very closely linked, then the possibility exists that virtually no recombination has occurred between them. Thus all individuals in the population will have the same genotype for both markers, and information is lost if one of the markers is deleted from the analysis.

With mid-density BeadChips that include 50,000–60,000 markers, generally only 75% are retained. With high-density BeadChips that include up to 800,000 markers, less than half are retained, with most markers deleted due to redundancy. The BovineHD BeadChip (Illumina Inc., http://www.illumina.com/Documents/products/datasheets/datasheet_bovineHD.pdf) includes 777,962 SNP, but only 311,725 markers were found to be useful for analysis (VanRaden *et al.*, 2013b; Weller *et al.*, 2013).

Deviation of Genotype Frequencies from Their Expectations

An additional factor that is generally checked is deviation from expected Hardy–Weinberg frequency of genotypes. Denoting the two marker alleles A and B, according to Hardy–Weinberg equilibrium the expected frequencies of the three genotypes AA, AB, and BB should be p^2, $2p(1-p)$, and $(1-p)^2$, where p is the frequency of allele A. There are several common reasons that cause genotype frequencies to diverge significantly from these expectations:

1. Incorrect scoring of genotypes. For example, it may be easier to score heterozygotes than homozygotes. This would result in an observed preponderance of heterozygotes, even though the marker is actually in Hardy–Weinberg equilibrium.
2. Unequal viability of genotypes. The most common situation is "recessive lethals." Human diseases such as cystic fibrosis and Tay–Sachs disease are examples. Unlike these diseases, which generally result in full-term pregnancies, many other recessive lethals result in very early-term abortions. Thus the only observed effect is the lack of homozygotes for the defective allele. An example in cattle is complex vertebral malformation, which generally results in early-term abortion, but in rare cases results in full-term malformed calves (Thomsen *et al.*, 2005). VanRaden *et al.* (2011b) used the lack homozygous haplotypes in dairy cattle as a test for the presence of recessive lethal alleles.
3. Sex chromosome location. If the marker is located on the section of the X chromosome that is not complementary to the Y chromosome, then there will be only one copy of the allele, and no heterozygotes will be found.
4. Copy number variation. If the gene is present in the genome in multiple copies and more than one copy is "called," then the observed frequency of heterozygotes will be higher than expected. For example, if two copies are detected and assumed to be the same marker, then a homozygote will be called only if both copies are homozygous for the same allele.

Despite all these factors, significant deviations from Hardy–Weinberg expectations for markers correctly assigned to autosomes with high call rates are quite rare.

Inclusion of All Markers versus Selection of Markers with Significant Effects

VanRaden *et al.* (2009) found that major reduction in the number of markers included in the analysis, compared to the approximately 40,000 markers retained, on the mid-density BeadChip affected accuracy of evaluation only slightly. Furthermore, there is general agreement that most markers do not have measurable effects on any specific quantitative traits. It therefore seemed logical to assume that appropriate selection of markers should increase the accuracy of evaluation or at least result in no reduction. In this case genotyping and analysis costs could be significantly reduced.

Various studies have proposed computation of GEBV based on subsets of SNPs. Four basic strategies have been proposed to select SNPs: random selection (Vazquez *et al.*, 2010); equally spaced SNPs throughout the genome (Habier *et al.*, 2009; VanRaden *et al.*, 2009; Weigel *et al.*, 2009; Moser *et al.*, 2010; Vazquez *et al.*, 2010; Zhang *et al.*, 2011; Weller *et al.*, 2014b); selection of SNPs with the greatest effects on the trait analyzed, as estimated from the analysis of all markers in the training set (Weigel *et al.*, 2009; Moser *et al.*, 2010; Vazquez *et al.*, 2010; Zhang *et al.*, 2011); and selection of markers based on principal component analysis (Pintus *et al.*, 2012). Accuracies nearly equal to analysis with all markers were obtained with subsets of markers, but in nearly all studies the accuracy of GEBV computed from subsets of markers was never significantly more than the accuracy of GEBV computed from analysis of all markers. An exception was Weller *et al.* (2014b) who selected markers based on their fixed additive effect on the bulls' current genetic evaluations for each trait, as derived by analysis of all valid SNPs by the "EMMAX" algorithm (Kang *et al.*, 2010). In this case the mean increase of the accuracy in the evaluations of young bulls between GEBV computed on selected subsets of 800 to 5000 markers and GEBV using all markers was 0.36. However, the proposed method requires information only available once the young bulls have daughters with records. A slight increase in accuracy was also obtained if markers were selected based on the change in their allele frequencies between young and old bulls.

Daetwyler *et al.* (2008) derived the following equation for the expected accuracy of the prediction of the additive genetic value ($r_{g\hat{g}}$) of an individual that can be achieved based on the measurement of n_P phenotypes assuming that n_G potential loci affect the trait of interest:

$$r_{g\hat{g}} = \sqrt{\frac{\lambda h^2}{\lambda h^2 + 1}} \tag{14.1}$$

where h^2 is the observed heritability and $\lambda = n_P/n_G$. In single-step analyses the "phenotypes" are DYD or deregressed genetic evaluations. In this case, the "heritability" is the mean reliability of the evaluations, which will generally be close to 0.9. The actual number of loci affecting the trait is of course unknown, but for the accuracy of prediction to equal 0.75 requires $\lambda = n_P/n_G = 1.43$. That is, the number of phenotypes should be approximately equal to 1.4 times the number of loci affecting the trait.

The Genomic Relationship Matrix

In Chapter 8 we introduced the numerator relationship matrix within the context of mixed model genetic evaluation. This is a symmetrical matrix with rows and columns equal to the number of animals included in the analysis. The elements of this matrix correspond to the fraction of genes identical by descent between the individuals represented by the corresponding row and column. For example, the element corresponding to the relationship between parent and offspring will generally have a value of 0.5. The diagonal elements of this matrix are generally equal to one, because an individual has all its genes in common with itself. However, for inbred individuals the diagonal value can be greater than one, because these individuals will have less genetic variation than outbred individuals.

Similar to the general relationship matrix, a genomic relationship matrix can be defined, in which elements will represent the overall covariances among the genotypes of individuals for all markers genotyped. All studies that have proposed methods for construction of the genomic relationship matrix have assumed that at each locus there are only two possible alleles, and the effects are additive, as described previously in this chapter.

Two caveats that must be considered are that the variance of each individual marker will be a function of the overall allelic frequencies, which will vary among markers, and that not all markers will have valid genotypes for all individuals. In addition the genomic variance matrix is generally divided by a constant, so that the values of the elements are comparable to the additive genetic matrix. Finally we should note that for genetic evaluation, the inverse of the relationship matrix is required. Unlike the additive numerator relationship matrix, which can be easily inverted by an algorithm developed by Henderson (1976), there is no known algorithm to derive the inverse of the genomic relationship matrix. This question will be considered further in the following chapter.

Summary

The basic concepts and strategies for genomic evaluation based on high-density SNP chips were discussed. We first considered the steps required to obtain genomic evaluations and the sources of bias in genomic evaluation and then considered the standard method for evaluation of genomic BV based on division of the population into training and validation subsets. We then dealt with the question of whether SNP effects should be considered fixed or random and explained why nearly all studies have assumed that SNP effects are random. We explained why nearly all algorithms for genomic evaluation have been based on analysis of individual SNPs, rather than haplotypes consisting of several SNPs, and explained the criteria used to select "useful" SNPs. Numerous studies have considered genomic evaluation based on subsets of SNPs, but in nearly all cases these result in a reduction in accuracy of evaluation. In the final sections we considered the basic questions required for construction of a genomic variance matrix. In the next chapter we will explain the most common methods for genomic evaluation in detail.

15 Genetic Evaluation Based on Analysis of Genetic Evaluations or Daughter-Yield Evaluations

Introduction

Methods to compute genomic genetic evaluations can be divided into two groups, denoted single-step methods and multistep methods. In single-step methods actual records are analyzed, and the effects of markers and the additive genetic variation not included in the marker effects are estimated. In multistep methods, discussed briefly in the previous chapter, genetic evaluations are first computed based on phenotypic data and pedigree. In the second step, the genetic evaluations of individuals with genotypes or a similar statistic is then analyzed as a function of the genetic markers. In the final step the direct genomic evaluations are generally incorporated into an index that includes pedigree information in addition to the direct genomic evaluations.

After a short discussion of the advantages and disadvantages of both types of analyses, this chapter will concentrate on methods proposed for multistep analysis, which are still the method of choice in most commercial evaluation systems. The following chapter will deal with single-step methods in detail and compare the results between the two types of methods.

Comparison of Single-Step and Multistep Models

As noted in the introduction, in single-step methods actual records are analyzed, and the effects of markers and the additive genetic variation not included in the marker effects are estimated. "Nuisance" effects such as herd-year-season or sex are also included in the model. If a single-trait animal model is employed, there will be an equation for the additive genetic effect of each animal, an equation for each marker, and equations for all levels of the fixed effects. Single-step methods have four drawbacks:

1. They require much more extensive computing, due to the generally huge number of equations included in the analysis model.
2. Single-step models must deal with the problem that the vast majority of animals with records do not have genotypes.
3. It is necessary to partition the genetic variance between the fraction associated with markers and the remainder that is independent of the marker effects. Although this factor can

Genomic Selection in Animals, First Edition. Joel Ira Weller.

theoretically be computed for historical data, it is unknown for animals that have not yet produced records.

4. Due to the huge numbers of equations, solutions can only be derived by iteration, and convergence to solutions can be a problem.

In multistep methods genetic evaluations are first computed based on phenotypic data and pedigree. In the second step, the daughter yield deviations (DYD) or a similar statistic is then analyzed as a function of the genetic markers. Derivation of DYD will be explained in the following section. Thus in the analysis model for the second step, the only effects included are the marker effects. Advantages of the multistep system for genomic evaluation include no change to the regular evaluations and simple steps for predicting genomic values for young genotyped animals. Furthermore, overall computing time will be considerably less than for single-step methods.

Disadvantages include weighting parameters, such as variance components (Guillaume *et al.*, 2008) or selection index coefficients (VanRaden *et al.*, 2009), loss of information, and biased evaluations (Misztal *et al.*, 2009; Aguilar *et al.*, 2010). Furthermore, the extension to alternative analysis models such as multitrait evaluations or test-day models is not obvious. Several problems exist in the use of DYD and yield deviations. These problems are weights (caused by different amount of information in the original data set), bias (e.g., caused by selection), accuracy (for animals in small herds), and collinearity (e.g., the yield deviations of two cows in the same herd). Furthermore, if genomic selection is used, the expectation of Mendelian sampling in selected animals is not zero, which can lead to biased evaluations (Patry and Ducrocq, 2010).

Although single-step methodologies appear to be superior on theoretical considerations, differences in accuracy of prediction between the two methods on actual data are at best minimal. These results will be discussed in more detail in the following chapter.

Derivation and Properties of Daughter Yields and DYD

"Daughter yields" are means of the records of a cow corrected for fixed effects, such as herd-year-seasons, but unlike genetic evaluations derived by mixed model methodology are not regressed toward the population means. Similarly, "DYD" are means of the records of the daughter of a sire corrected for fixed effects and also for merit of mates and account for unequal numbers of records among cows. Like daughter yields, DYD are not regressed toward the population mean as a function of the reliability of the evaluation. These statistics were originally derived as by-products of genetic evaluations derived by the individual animal model (VanRaden and Wiggans, 1991).

In section "Important Properties of Mixed Model Solutions" of Chapter 6, we explained that the ratio Var($\hat{\mathbf{u}}$)/Var($\mathbf{u}$) is called the "reliability" of $\mathbf{u}$, where $\mathbf{u}$ is the actual breeding value and $\hat{\mathbf{u}}$ is the estimated breeding value. The reliability is theoretically equal to the square of the correlation between $\hat{\mathbf{u}}$ and $\mathbf{u}$, that is, the coefficient of determination. For genetic evaluations derived by mixed model methodology, the variance of the evaluation increases, and the prediction error variance decreases as functions of the amount of information on which the evaluation is based, as demonstrated by Equation (6.20), which will be repeated here:

$$\mathrm{Var}(\mathbf{u}) = \mathrm{Var}(\hat{\mathbf{u}}) + \mathrm{PEV}(\hat{\mathbf{u}}) \tag{15.1}$$

where PEV($\hat{\mathbf{u}}$) is the prediction error variance of $\hat{\mathbf{u}}$ which is equal to the corresponding diagonal elements of the inverse of the coefficient matrix. Although exact computation of the variances of the genetic evaluations requires inversion of the coefficient matrix, algorithms to derive approximate values have been derived which do not require inversion of this very large matrix (Misztal and Wiggans, 1988; Misztal *et al.*, 1991).

This is not the case for simple means, least squares means, yield deviations, or DYD. Variances of simple means and DYD decrease as the number of records on which the DYD is based increases. In a simple least squares analysis, the residual variance of the dependent variable is considered equal for all records. In a generalized least squares analysis, records are weighted by the inverse of the residual variance matrix. That is, records with lower residual variances (i.e., records based on more data) are given greater weight. This is reasonable if records with more data do have lower variances, which is the case for DYD, but not genetic evaluations, as explained previously.

For this reason, for genomic evaluation by two-step methodologies, the dependent variable is generally the DYD of the trait analyzed if sires are genotyped or yield deviations if cows are genotyped. Although DYD should then be weighted by their variances, these variances are difficult to compute, and DYD are usually weighted by some function of the reliability of the evaluations, as will be explained in the following section. It is assumed that this function is approximately proportional to the DYD variances.

Computation of "Deregressed" Genetic Evaluations

DYD are computed only for traits that are analyzed by a standard single-trait animal model. Thus DYD are generally not computed if some other model is used for genetic evaluation. Also even if the basic traits are analyzed by a standard animal model, genetic evaluations for some traits are computed as function of the genetic evaluations for the basic traits. For example, fat and protein concentration are computed as functions of fat, protein, and milk yield; selection indices are generally computed as linear functions of several traits. In all of these cases "deregressed" genetic evaluations are analyzed, instead of the DYD. Generally genetic evaluations are deregressed by division by a simple function of the reliability, DRP, as proposed by VanRaden *et al.* (2009):

$$\mathrm{DRP} = \mathrm{PA} + \left(\mathrm{EBV} - \mathrm{PA}\right) * \left(\frac{\mathrm{EDC}_{\mathrm{parents+progeny}}}{\mathrm{EDC}_{\mathrm{progeny}}}\right) \tag{15.2}$$

where EBV = estimated breeding value, PA = parent average EBV (in case the dam EBV has not been computed; the PA is replaced by the sire pedigree index, which is equal to ½*(sire EBV) + ¼*(maternal grandsire EBV)); and EDC = estimated daughter contributions. For each offspring i with a record, of the sire, EDC_i is computed as follows:

$$\mathrm{EDC}_i = \frac{k\left(\mathrm{rel}_i\right)}{4 - \mathrm{rel}_i\left(1 + \mathrm{rel}_{\mathrm{d}}\right)} \tag{15.3}$$

where rel_i and $\mathrm{rel}_{\mathrm{d}}$ are the reliabilities of the offspring and her dam, and $k = (4 - h^2)/h^2$, where h^2 is the heritability. $\mathrm{EDC}_{\mathrm{parents+progeny}}$ is then computed as the sum of EDC_i over all offspring of the sire with records. Alternatively, $\text{EDC}_{\text{parents+progeny}}$ can be estimated from the reliabilities of EBV of the sires as follows (Přibyl *et al.*, 2013):

$$\text{EDC}_{\text{parents+progeny}} = \frac{\left(1-0.25h^2\right)\text{rel}_{\text{s}}}{0.25h^2\left(1-\text{rel}_{\text{s}}\right)} \tag{15.4}$$

where rel_{s} is the reliability of the sire. $\text{EDC}_{\text{parents}}$ is computed in the same way, except that the reliability of the sire is replaced by the reliability of the PA. $\text{EDC}_{\text{progeny}}$ is then computed as follows:

$$\text{EDC}_{\text{progeny}} = \text{EDC}_{\text{parents+progeny}} - \text{EDC}_{\text{parents}} \tag{15.5}$$

For sires with hundreds of daughters, $\text{EDC}_{\text{parents}}$ will be negligible compared to $\text{EDC}_{\text{parents+progeny}}$; $\text{EDC}_{\text{progeny}} \approx \text{EDC}_{\text{parents+progeny}}$, and DRP ≈ EBV. For sires with relatively few progeny, $\text{EDC}_{\text{parents+progeny}}/\text{EDC}_{\text{progeny}} > 1$.

Analysis of DYD as the Dependent Variable with All Markers Included as Random Effects

This method was first proposed by VanRaden (2008). If each individual is measured once for a trait and the inheritance of all alleles is known, then data vector **y** can be modeled as a general mixed model similar to Equation (6.14) and now repeated:

$$\mathbf{y} = \mathbf{xb} + \mathbf{Zu} + \mathbf{e} \tag{15.6}$$

where **b** is the mean (the only fixed effect in the model); **x** is a vector of 1's; **u** are the marker effects, which are assumed to be random; **Z** is the matrix that relates marker effects to the individual records; and **e** is the random error vector with variance matrix **R**. As noted in the previous section, this model requires weighting the residuals as a function of the DYD reliabilities. Diagonals of **R** were computed as $(1/(\text{R}_{\text{dau}}-1)\sigma_{\text{e}}^2)$, where R_{dau} is the bull's reliability from daughters with parent information excluded and σ_{e}^2 is the residual variance of the DYD not explained by the marker effects. All markers were assumed to be biallelic.

Let **M** be the matrix that specifies which marker alleles each individual inherited. Dimensions of **M** are the number of individuals by the number of markers. If elements of **M** are set to −1, 0, and 1 for the homozygote, heterozygote, and other homozygote, respectively, diagonals of **MM′** count the number of homozygous loci for each individual, and off-diagonals measure the number of alleles shared by relatives. Let the frequency of the second allele at locus i be p_i, and let the matrix **P** contain allele frequencies expressed as a difference from 0.5 and multiplied by 2, so that column i of **P** is $2(p_i-0.5)$. **Z** is then defined as **M** − **P**, so that mean values of the allele effects in **Z** = 0. VanRaden (2008) assumed that allelic frequencies would be computed from the "base animals," that is, animals with genotypes, but without ancestors with genotypes. Aguilar *et al.* (2010) investigated this question for single-step models and concluded that assuming $p_i=0.5$ resulted in optimal genomic evaluations. This question will be considered again in Chapter 17.

VanRaden (2008) gives three different methods to derive genomic breeding values. In the first method the effects of the individual markers can be derived from the following equations:

$$\left[\mathbf{Z}'\mathbf{R}^{-1}\mathbf{Z}+\mathbf{I}\lambda\right]\left[\mathbf{u}\right]=\mathbf{Z}'\mathbf{R}^{-1}\left(\mathbf{y}-\mathbf{x}\hat{\mathbf{b}}\right) \tag{15.7}$$

where λ = the ratio σ_e^2/σ_u^2, which equals the sum across marker loci $2\Sigma p_i(1-p_i)$ times the ratio σ_e^2/σ_a^2, where σ_a^2 is the total genetic variance and $\mathbf{x}\hat{\mathbf{b}}$ are the solutions for the means of $\mathbf{y}$. Genomic breeding values are then obtained as $\mathbf{Zu}$^. As noted previously, $\mathbf{R}$ is a diagonal matrix; therefore $\mathbf{R}^{-1}$ is computed by inverting each diagonal element. Thus this method is computationally tractable, as it is not necessary to invert any large matrices, and $\mathbf{u}$ can be solved by standard iteration algorithms, such as Gauss–Seidel. However, iteration will be computing intensive, because, unlike standard mixed model equations, all elements will have nonzero values.

In the second method it is first necessary to compute the "genomic relationship matrix," $\mathbf{G}$. Similar to the additive genetic relationship matrix, this matrix describes the covariance among individuals, but in this case the covariance is computed relative to the sum of similarities among the marker genotypes. This matrix can be derived as follows:

$$\mathbf{G}=\frac{\mathbf{ZZ}'}{2\Sigma p_i\left(1-p_i\right)} \tag{15.8}$$

Division by $2\Sigma p_i(1-p_i)$ scales $\mathbf{G}$ to be analogous to the numerator relationship matrix $\mathbf{A}$. Genomic estimated breeding values (GEBV) can then be derived by the selection index equation

$$\mathbf{u}=\mathrm{E}\left(\mathbf{u}\right)+\mathbf{CV}^{-1}\left[\mathbf{y}-\mathrm{E}\left(\mathbf{y}\right)\right] \tag{15.9}$$

where $\mathbf{u}$ is the vector of estimated genetic values, $\mathbf{C}$ is the covariance matrix between $\mathbf{u}$ and $\mathbf{y}$, and $\mathbf{V}$ is the variance matrix of $\mathbf{y}$. In this case, since DYD are analyzed and no fixed effects are included in the model, E($\mathbf{u}$) can be deleted. If DYD and genotypes are available on all individuals included in the analysis, $\mathbf{C}=\mathbf{G}$, and $\mathbf{V}=\mathbf{G}+\mathbf{R}(\sigma_e^2/\sigma_a^2)$, where σ_a^2 is total additive genetic variance. Thus the GEBV can be computed as follows:

$$\hat{\mathbf{a}}=\mathbf{G}\left[\mathbf{G}+\mathbf{R}\left(\frac{\sigma_e^2}{\sigma_a^2}\right)\right]^{-1}\left(\mathbf{y}-\mathbf{X}\hat{\mathbf{b}}\right) \tag{15.10}$$

where $\hat{\mathbf{a}}$ are the estimated GEBV and $\mathbf{X}\hat{\mathbf{b}}$ are the solutions for the means of $\mathbf{y}$, the vector of DYD. Note that the solution of this set of equations does not compute solutions for the individual marker effects. Also this method requires inversion of a matrix of rank equal to the number of animals included in the analysis. All elements of $\mathbf{G}$ will be nonzero; thus sparse matrix techniques cannot be applied to compute the inverse.

A third solution strategy presented by Garrick (2007) could be more efficient than the selection index, because $\mathbf{G}$ can be inverted just once and then additional traits with differing heritability or $\mathbf{R}$ processed using iteration:

$$\hat{\mathbf{a}}=\left[\mathbf{R}^{-1}+\mathbf{G}^{-1}\left(\frac{\sigma_e^2}{\sigma_a^2}\right)\right]^{-1}\mathbf{R}^{-1}\left(\mathbf{y}-\mathbf{X}\hat{\mathbf{b}}\right) \tag{15.11}$$

The matrix **G** may be singular, for example, if the number of markers does not exceed the number of individuals genotyped, but this will generally not be a problem.

Final GEBV are derived by the selection index as shown in Equation (13.1) which includes three factors: (i) the estimated direct genomic evaluations as described, (ii) the parent average EBV computed from the subset of genotyped ancestors using known relationships, and (iii) the parent average EBV for individuals with EBV for both parents, or pedigree indices of individuals without dam EBV.

Computation of Reliabilities for Genomic Estimated Breeding Values

In section "Important Properties of Mixed Model Solutions" of Chapter 6, we explained that reliabilities of genetic evaluations are computed as Var($\hat{\mathbf{u}}$)/Var($\mathbf{u}$) and that Var($\mathbf{u}$) = Var($\hat{\mathbf{u}}$) + PEV($\hat{\mathbf{u}}$), where PEV($\hat{\mathbf{u}}$) is the prediction error variance of $\hat{\mathbf{u}}$ which are equal to the corresponding diagonal elements of the inverse of the coefficient matrix. Reliabilities of GEBV for bulls with DYD were obtained from

$$\text{Diag}\left\{\mathbf{G}\left[\mathbf{G}+\mathbf{R}\left(\frac{\sigma_e^2}{\sigma_a^2}\right)\right]^{-1}\mathbf{G}'\right\} \qquad (15.12)$$

where Diag{.} are the diagonal elements of the matrix. As in Equation (15.10) computation of the reliabilities requires inversion of the matrix $\mathbf{G}+\mathbf{R}\left(\sigma_e^2/\sigma_a^2\right)$. Reliabilities obtained by this expression were compared to the bull reliabilities obtained by standard animal model evaluations. VanRaden *et al.* (2009) used the increase in reliability to evaluate the expected increase in genetic gain due to marker information.

Bayesian Weighting of Marker Effects

In Chapter 7 we described application of Bayesian methodology for a granddaughter design analysis. In Bayesian estimation of QTL effects, data on the prior distribution of QTL effects is included in the estimation of the effects. That is, smaller effects are regressed more toward the mean as compared to larger effects. In application of Bayesian methodology to genomic evaluation, three basic models have been proposed, denoted "Bayes A," "Bayes B," and "Bayes C." Bayes A models assume a continuous prior distribution of QTL effects with a nonzero effect for all comparisons tested, while in "Bayes B" models a zero effect is assumed for the majority of the markers genotyped (Meuwissen *et al.*, 2001). Although Bayes A models are easier to apply, Bayes B models are closer to reality. Kizilkaya *et al.* (2010) proposed a Bayes C model that assumes a common variance for all makers with nonzero effects estimated from the data, instead of the locus-specific variance in Bayes B. Habier *et al.* (2011) proposed a modified Bayes C model, denoted "Bayes-Cπ," by treating the probability π that a marker has an effect as an unknown parameter, which is estimated from the data.

In the Bayes A analysis of VanRaden (2008), he assumed that the prior distribution was a simple, heavy-tailed distribution generated from a normal variable divided by $1.25^{\text{abs}(s-2)}$, where s is the number of standard deviations from the mean and 1.25 determines departure from normality. Defining $\lambda=\sigma_e^2/\sigma_a^2$, the constant value of λ for all markers in Equation (15.7) is replaced by individual λ_i for each marker computed as $\lambda_i=\lambda/1.25^{\text{abs}(s-2)}$. Unlike the model of Weller *et al.* (2005), the marker effect in this model can be either positive or negative.

In the Bayes B analysis, VanRaden (2008) assumed that only 700 markers of 50,000 included in the analysis have nonzero effects. In this case λ_i was computed as follows:

$$\lambda_i = \lambda\left[\frac{q}{m} + \left(1 - \frac{q}{m}\right)\left(\frac{f_{\text{err}}}{f_{\text{QTL+err}}}\right)\right] \tag{15.13}$$

where *q/m* is the fraction of markers assumed to have effects on the trait analyzed, f_{err} is the density function for those markers that do not have effects, and $f_{\text{QTL+err}}$ is the density function for those markers that do have effects on the trait analyzed.

Additional Bayesian Methods for Genomic Evaluation

The linear method of VanRaden (2008) is basically a BLUP method in which additional weight is put on the diagonals relative to off-diagonal elements of the coefficient matrix. Thus, unlike simple least squares estimate, there is no problem of "overparameterization" even though the number of markers is greater than the number of individuals with DYD. Various other penalized regression methods have been proposed, for example, least absolute shrinkage and selection operator (LASSO) (Tibshirani, 1996), which adds $\mathbf{l}_1$ penalty functions to the traditional least squares. LASSO and its extensions including elastic net (Zou and Hastie, 2005) and adaptive LASSO (Zou, 2006) have been used in various QTL mapping or genomic selection studies (de los Campos *et al.*, 2009; Legarra *et al.*, 2011).

None of these two-step Bayesian methods have been demonstrated to be significantly superior to the original method of VanRaden (2008), based on their ability to predict breeding values of young bulls with genotypes, but without daughter records. As will be seen in the following chapter, this is also the case for single-step methods.

Ober *et al.* (2011) proposed a nonparametric kernel-based method, which they tested on simulated data, assuming that all individuals of the training population had both phenotypes and genotypes. In the presence of dominance or epistasis for the QTL effects, their method was able to slightly outperform the method of VanRaden (2008).

Summary

The huge number of marker genotypes available in modern BeadChips, in the range of 50,000–60,000 for "moderate-density" chips, and the fact that only a small fraction of the population with records will be genotyped created new problems that were generally not considered in traditional methods for genetic evaluations. To solve these problems, two general strategies have been proposed to derive genomic genetic evaluations: Analysis of only individuals with genotypes, in this case the dependent variable is a function of the genetic evaluations based only on phenotypic data, and inclusion of marker data together with phenotypic data in a single-step evaluation. The former method was denoted "multistep" evaluation and was discussed in detail. We also considered the proposed variations of this method based on several different Bayesian models. However, none of these models are able so far to significantly increase the accuracy of genomic evaluations. In the following chapter single-step methods will be described in detail.

16 Genomic Evaluation Based on Analysis of Production Records

Introduction

In the previous chapter we explained that methods to compute genomic genetic evaluations can be divided into two groups, denoted single-step methods and multi- or two-step methods. Multistep methods, in which genetic evaluations are first computed based on phenotypic data, were described in detail in Chapter 15. In this chapter we will describe single-step methods in detail. In these methodologies, the phenotypic records are the dependent variables, and the analysis model includes the polygenic effect, the effects of the individual markers, and all fixed effects, such as herd-year-season.

The major difficulties with application of single-step methodologies derive from the fact that generally only a small fraction of the population will be genotyped. Furthermore, the genotyped individuals will generally be males without records for traits related to female fertility and milk production. Because of the generally huge number of records, overparameterization of the model is no longer a problem, even if marker effects are considered fixed. However, nearly all single-step studies have also assumed that marker effects are random and algorithms are based on Henderson's mixed model equations. Therefore we will first briefly review Henderson's mixed model equations and then discuss the modifications required for single-step genomic evaluation.

Single-Step Methodologies: The Basic Strategy

In two-step methodology as described in Equations (15.6) through (15.11), the dependent variables were a function of the genotyped animals' estimated breeding values, and the model effects were the individual effects of each marker. Genomic evaluations were then computed by summing the individual marker effects. In single-step methods, similar to the second method proposed by VanRaden (2008), the effects of the individual markers are incorporated through computation of a "genomic relationship matrix." Computation of this matrix is given in Equation (15.8) and was described in detail in the previous chapter. Upon division by a constant, the scale of this matrix can be made equal to the numerator relationship matrix, **A**. If all individuals included in the analysis were genotyped and if the genomic variance matrix accounts for all additive genetic variance, it should only be necessary to replace the inverse of **A** with the inverse of the genomic relationship matrix. However, this is problematic for two reasons. First, as noted previously, only a small fraction

Genomic Selection in Animals, First Edition. Joel Ira Weller.

of the population will generally have genotypes; and unlike the standard relationship matrix, there is no simple algorithm to compute the inverse of the genomic relationship matrix. Both of these problems have been solved, and the solutions will be described in the following two sections.

Computation of the Modified Relationship Matrix when only a Fraction of the Animals are Genotyped: The Problem

At first thought it would seem that it should be possible to use the algorithm of VanRaden (2008) given in Equation (15.8) to compute the genomic relationships among animals with genotypes and the standard rules of Mendelian inheritance to compute relationships among nongenotyped animals and between genotyped and nongenotyped animals. Assume that the numerator relationship matrix is partitioned as follows:

$$\mathbf{A} = \begin{bmatrix} \mathbf{A}_{11} & \mathbf{A}_{21} \\ \mathbf{A}_{12} & \mathbf{A}_{22} \end{bmatrix} \tag{16.1}$$

where $\mathbf{A}_{11}$ refers to additive genetic relationships among animals without genotypes, $\mathbf{A}_{22}$ refers to genomic relationships among genotyped animals, and $\mathbf{A}_{12}$ and $\mathbf{A}_{21}$ refer to additive genetic relationships between genotyped and ungenotyped individuals. Since the matrix is symmetrical, $\mathbf{A}_{12} = \mathbf{A}'_{21}$. In the first instance we will assume that $\mathbf{A}_{22}$ is replaced by $\mathbf{G}$ as computed by the method of VanRaden (2008). Then $\mathbf{A}_g$, the modified relationship matrix that accounts for relationships due to markers, is computed as follows:

$$\mathbf{A}_g = \begin{bmatrix} \mathbf{A}_{11} & \mathbf{A}_{21} \\ \mathbf{A}_{12} & \mathbf{G} \end{bmatrix} \tag{16.2}$$

Criteria for Valid Genetic Relationship Matrices

We will now explain why this solution cannot be applied in practice. Symmetric matrices can be divided into five groups: positive definite, negative definite, positive semidefinite, negative semidefinite, and indefinite. Assume a symmetric matrix $\mathbf{C}$ and a vector $\mathbf{x}$ of length equal to the dimensions of $\mathbf{C}$, then $\mathbf{C}$ is "positive definite" if $\mathbf{x}'\mathbf{C}\mathbf{x}>0$ for any vector $\mathbf{x}$, other than $\mathbf{x}=0$. Similarly if $\mathbf{x}'\mathbf{C}\mathbf{x}<0$ for any vector $\mathbf{x}$, other than $\mathbf{x}=0$, then the matrix is "negative definite." If $\mathbf{x}'\mathbf{C}\mathbf{x}\geq 0$ for any vector $\mathbf{x}$, other than $\mathbf{x}=0$, then $\mathbf{C}$ is "positive semidefinite," and if $\mathbf{x}'\mathbf{C}\mathbf{x}\leq 0$ for any vector $\mathbf{x}$, other than $\mathbf{x}=0$, then $\mathbf{C}$ is "negative semidefinite." If $\mathbf{x}'\mathbf{C}\mathbf{x}>0$ for some vector $\mathbf{x}$, but $\mathbf{y}'\mathbf{C}\mathbf{y}<0$ for some vector $\mathbf{y}$, then the matrix is "indefinite." All eigenvalues of a positive definite matrix will be positive. If the matrix is positive semidefinite, then some eigenvalues will be zero. For an indefinite matrix, or a negative semidefinite matrix, some eigenvalues will be negative. Neither matrix can be a valid relationship matrix. This could potentially result in discrepancies that contradict the principles of quantitative genetics, as will be seen in the following example.

As shown by Legarra *et al.* (2009), the use of $\mathbf{G}$ potentially modifies covariances in ancestors and descendants of genotyped animals. For example, assume two full-sibs in the genotyped animals whose genomic relationship is 0.6. By using $\mathbf{A}_g$, it is assumed that the average relationship among their daughters is 0.25, whereas in fact it is 0.3. It can be verified by small numerical examples that $\mathbf{A}_g$ is indefinite (i.e., some eigenvalues are negative and some are positive).

Computation of the Modified Relationship Matrix when only a Fraction of the Animals are Genotyped, the Solution

Legarra *et al.* (2009) derived the following relationships for a population that includes both genotyped and ungenotyped individuals:

$$\mathrm{Var}(\mathbf{u}_1) = \mathbf{A}_{11} + \mathbf{A}_{12}\mathbf{A}_{22}^{-1}(\mathbf{G} - \mathbf{A}_{22})\mathbf{A}_{22}^{-1}\mathbf{A}_{21} \tag{16.3}$$

$$\mathrm{Var}(\mathbf{u}_2) = \mathbf{G} \tag{16.4}$$

$$\mathrm{Cov}(\mathbf{u}_1,\mathbf{u}_2) = \mathbf{A}_{12}\mathbf{A}_{22}^{-1}\mathbf{G} \tag{16.5}$$

where $\mathbf{u}_1$ and $\mathbf{u}_2$ are the breeding values for ungenotyped and genotyped individuals, respectively, and $\mathbf{G}$ is the genomic variance matrix as given in Equation (15.8). Note that in this case $\mathbf{G} \neq \mathbf{A}_{22}$. Based on these relationships, Legarra *et al.* (2009) define $\mathbf{H}$, the covariance matrix of breeding values including genomic information as follows.

$$\mathbf{H} = \begin{bmatrix} \mathbf{A}_{11} + \mathbf{A}_{12}\mathbf{A}_{22}^{-1}(\mathbf{G} - \mathbf{A}_{22})\mathbf{A}_{22}^{-1}\mathbf{A}_{21} & \mathbf{A}_{12}\mathbf{A}_{22}^{-1}\mathbf{G} \\ \mathbf{G}\mathbf{A}_{22}^{-1}\mathbf{A}_{21} & \mathbf{G} \end{bmatrix} \tag{16.6}$$

Note that all elements of these equations other than $\mathbf{G}$ are either submatrices of $\mathbf{A}$ or of $\mathbf{A}^{-1}$, which can all be readily computed. However, solving the standard mixed model equations requires the inverse of $\mathbf{H}$, which cannot be inverted in real time by standard inversion algorithms for populations of commercial size. Two solutions to solving the mixed model equations will be presented in the following two sections.

Solving the Mixed Model Equations without Inverting H

Replacing the standard numerator relationship matrix with $\mathbf{H}$ in the standard mixed model equations given in Equation (6.17) and ignoring the permanent environmental effect give the following system of equations:

$$\begin{bmatrix} \mathbf{X}'\mathbf{X} & \mathbf{X}'\mathbf{Z} \\ \mathbf{Z}'\mathbf{X} & \mathbf{Z}'\mathbf{Z} + \mathbf{H}^{-1}\alpha \end{bmatrix} \begin{bmatrix} \hat{\boldsymbol{\beta}} \\ \hat{\mathbf{u}} \end{bmatrix} = \begin{bmatrix} \mathbf{X}'\mathbf{y} \\ \mathbf{Z}'\mathbf{y} \end{bmatrix} \tag{16.7}$$

where $\alpha = \sigma_e^2/\sigma_g^2$, where σ_g^2 is the variance accounted for by the genetic markers. Misztal *et al.* (2009) proposed multiplying the second set of equations by $\mathbf{H}^{-1}$ to yield the following set of equations, which include $\mathbf{H}$, but not $\mathbf{H}^{-1}$:

$$\begin{bmatrix} \mathbf{X}'\mathbf{X} & \mathbf{X}'\mathbf{Z} \\ \mathbf{H}\mathbf{Z}'\mathbf{X} & \mathbf{H}\mathbf{Z}'\mathbf{Z} + \mathbf{I}\alpha \end{bmatrix} \begin{bmatrix} \hat{\boldsymbol{\beta}} \\ \hat{\mathbf{u}} \end{bmatrix} = \begin{bmatrix} \mathbf{X}'\mathbf{y} \\ \mathbf{H}\mathbf{Z}'\mathbf{y} \end{bmatrix} \tag{16.8}$$

This system of equations is not symmetric, and the matrix $\mathbf{H}$ may be indefinite. Therefore standard methodologies such as Gauss–Seidel iteration may not converge. The standard algorithm for solving sparse systems with a nonsymmetric coefficient matrix is biconjugate gradient stabilized (Bi-CGSTAB; Van der Vorst, 1992). This algorithm requires storing twice the number of elements in the coefficient matrix times a vector product per round of iteration.

Inverting the Genomic Relationship Matrix

Although there is no simple algorithm to invert **H**, the inverse can be computed as follows (Aguilar *et al.*, 2010):

$$\mathbf{H}^{-1} = \mathbf{A}^{-1} + \begin{bmatrix} 0 & 0 \\ 0 & \mathbf{G}^{-1} - \mathbf{A}_{22}{}^{-1} \end{bmatrix} \tag{16.9}$$

However, **G** is usually singular, due to collinearity among marker genotypes and, therefore, cannot be inverted without additional steps. In the previous section we assumed that all genetic variance of the genotyped individuals can be explained by the genomic relationship matrix. This is clearly not the case if the matrix is derived from a mid-density SNP chip of approximately 50,000 markers. The problem of singularity can be solved by partitioning the genetic variance for the genotyped animals into a fraction explained by markers and an unexplained fraction as follows:

$$\mathbf{H}^{-1} = \mathbf{A}^{-1} + \begin{bmatrix} 0 & 0 \\ 0 & \lambda\left(\mathbf{G}^{-1} - \mathbf{A}_{22}^{-1}\right) \end{bmatrix} \tag{16.10}$$

where λ scales differences between genomic and pedigree-based information. Equations require the inverses of **A**, $\mathbf{A}_{22}$, and **G**. The first two can be computed by the algorithm of Henderson (1976). The inverse of **G**, which included only genotyped individuals, will have a rank of several thousand and can therefore be computed in real time by standard algorithms for matrix inversion. A drawback of this method is that there is no method to accurately estimate λ. Aguilar *et al.* (2010) proposed testing a range of values on a test data set.

Misztal *et al.* (2010) proposed an additional modification to Equation (16.10) as follows:

$$\mathbf{H}^{-1} = \mathbf{A}^{-1} + \begin{bmatrix} 0 & 0 \\ 0 & \tau\mathbf{G}^{-1} - \omega\mathbf{A}_{22}^{-1} \end{bmatrix} \tag{16.11}$$

where τ is a scale factor for $\mathbf{G}^{-1}$ and ω is a scale factor for $\mathbf{A}_{22}^{-1}$. Changes in both weights were investigated for final score for US Holsteins. While changes in τ had little influence on the accuracy and bias of prediction, values smaller than 1 for ω helped to reduce inflation of GEBV. Tsuruta *et al.* (2011) proposed values of 1.0 for τ and 0.7 for ω for analysis of linear type traits, but again these conclusions are based on optimization of results for a specific data set.

Estimation of Reliabilities for Genomic Breeding Values Derived by Single-Step Methodologies

In section "Important Properties of Mixed Model Solutions" of Chapter 6, we explained that reliabilities of genetic evaluations are computed as $\text{Var}(\hat{\mathbf{u}})/\text{Var}(\mathbf{u})$ and that $\text{Var}(\mathbf{u}) = \text{Var}(\hat{\mathbf{u}}) + \text{PEV}(\hat{\mathbf{u}})$, where $\text{PEV}(\hat{\mathbf{u}})$ is the prediction error variance of $\hat{\mathbf{u}}$ which are equal to the corresponding diagonal elements of the inverse of the coefficient matrix. In the previous chapter we showed that reliabilities for the multistep methodology can be computed from Equation (15.12). These equations require inversion of a matrix with rank equal to the number of animals included in the analysis. This is possible for multistep methodologies in which only the sires with genotypes are included in the analysis, but is not a viable option for single-step methodologies, in which millions of animals are analyzed. Thus approximations have been derived which do not require complete inversion of the coefficient matrix.

A first approximation proposed by Ufford *et al.* (1979) was to invert the diagonal elements of the coefficient referring to animals. This method gave reasonable approximations for a sire model without relationships, but not for sire models with relationships or animals. Misztal and Wiggans (1988) proposed that for animal models reliabilities could be approximated as $1-[\alpha/(\alpha+d_i)]$, where α is the ratio of error variance to animal genetic variance and d_i is the amount of information for animals in units of effective number of records. For single-step evaluations, d_i can be partitioned as $d_i^{\mathrm{r}}+d_i^{\mathrm{p}}+d_i^{\mathrm{g}}$ where d_i^{r} is the contribution from records (phenotypes), d_i^{p} is the contribution from pedigrees, and d_i^{g} is the contribution from genomic information. The diagonal elements of the inverse of the coefficient matrix for animals can be computed as follows:

$$\mathrm{CM}^{ii}=\frac{1}{\alpha+d_i^{\mathrm{r}}+d_i^{\mathrm{p}}+d_i^{\mathrm{g}}} \tag{16.12}$$

where CM^{ii} is the diagonal element of the coefficient matrix referring to animal i. Misztal *et al.* (2013) proposed that CM^{ii} can be approximated by the following formula:

$$\mathrm{CM}^{ii}\approx\left\{\left[\mathbf{D}_i^{\mathrm{r}}+\mathbf{D}_i^{\mathrm{p}}+\left(\mathbf{I}+\mathbf{G}^{-1}-\mathbf{A}_{22}^{-1}\alpha\right)\right]^{-1}\right\}_{ii} \tag{16.13}$$

where $\mathbf{D}^{\mathrm{r}}=\left\{d_i^{\mathrm{r}}\right\}$ and $\mathbf{D}^{\mathrm{p}}=\left\{d_i^{\mathrm{p}}\right\}$. In this equation, $\mathbf{G}$ accounts for genomic information, and $\mathbf{A}_{22}$ accounts for an adjustment to prevent double counting of the relationship information contained in $\mathbf{G}$ and $\mathbf{A}$. Misztal *et al.* (2013) presented an algorithm to compute approximate reliabilities based on Equation (16.13). Correlations of the approximate reliabilities derived from this algorithm and actual reliabilities derived from the inverse of a simulated data set were 0.98.

Single-Step Computation of Genomic Evaluations with Unequally Weighted Marker Effects

A deficiency of the single-step algorithm described previously is that, unlike the multistep methodology presented in Equation (15.7), effects are not computed for individual markers. Rather the model is based only on overall marker identity among individuals, without regard to which chromosomal regions actually contain segregating QTL of interest. Wang *et al.* (2012) proposed a single-step algorithm that iteratively assigns weights to all markers, based on their solutions from the previous round of iteration. The animal breeding values are divided into those for genotyped ($\mathbf{a}_{\mathrm{g}}$) and ungenotyped ($\mathbf{a}_{\mathrm{n}}$) animals. As in the previous chapter, genomic breeding values are computed as $\mathbf{Zu}$, where $\mathbf{u}$ is the vector of marker effects and $\mathbf{Z}$ is the matrix relating marker effects to animals. The variance of $\mathbf{Zu}$ is computed as follows:

$$\mathrm{Var}(\mathbf{Zu})=\mathbf{ZDZ}'\sigma_{\mathrm{u}}^2=\mathbf{G}^*\sigma_{\mathrm{a}}^2 \tag{16.14}$$

where **D** is a diagonal matrix of weights for variances of SNP, σ_a^2 is the total genetic additive, σ_u^2 is the variance of the distribution for marker effects assuming they are identically distributed, and **G***
is a weighted genomic relationship matrix. The *i*th diagonal element of **D**, d_i, is computed as follows:

$$d_i = \hat{u}_i^2 2p_i(1-p_i) \quad (16.15)$$

where $\hat{u}_i^2$ is the solution for the *i*th marker effect and p_i is the allele frequency of the second allele of the *i*th marker. The joint (co)variance of $\mathbf{a}_g$ and **u** is computed as follows:

$$\mathbf{Var}\begin{bmatrix}\mathbf{a}_g \\ \mathbf{u}\end{bmatrix} = \begin{bmatrix}\mathbf{ZDZ'} & \mathbf{ZD'} \\ \mathbf{DZ'} & \mathbf{D}\end{bmatrix}\sigma_u^2 \quad (16.16)$$

Then **G*** can be computed as follows:

$$\mathbf{G^*} = \mathbf{ZDZ'}\frac{\sigma_u^2}{\sigma_a^2} = \frac{\mathbf{ZDZ'}}{\sum_{i=1}^{M} 2p_i(1-p_i)} \quad (16.17)$$

where, as in the previous chapter, *M* is the number of markers. Then, based on the selection index theory, solutions for the marker effects can be derived as follows:

$$\hat{\mathbf{u}} = \frac{\sigma_u^2}{\sigma_a^2}\mathbf{DZ'G^{*-1}\hat{a}_g} \quad (16.18)$$

Since the terms on the right-hand side of Equation (16.18) are themselves functions of $\hat{\mathbf{u}}$, these equations can only be solved iteratively. New values for $\hat{\mathbf{u}}$ and weights are obtained in each round.

The number of iterations needed is empirical and data dependent. The most appropriate weights for markers were obtained after one round of iteration on a moderately sized data set of approximately 700,000 cows (Lourenco *et al.*, 2014a). In most cases, some shrinkage of effects was obtained after the first round, and markers with very small effects had weights reduced to zero.

Summary

In single-step methodologies, genomic breeding values are derived by solving modified mixed model equations in which the independent variables are the actual animal records and the dependent variables are fixed effects—the sum of additive genetic effects explained by genetic markers and the additive genetic effects not explained by markers. Although the proposed methods required much more computing resources than two-step methods, genomic breeding values can be computed even for large commercial populations consisting of millions of records. Accurate approximate reliabilities can also be computed for single-step methods, and these methods can also be modified to selectively weight markers with greater effects on the trait analyzed.

17 Validation of Methods for Genomic Estimated Breeding Values

Introduction

Nearly all studies that have proposed methods for genomic evaluations have also attempted to validate the proposed methodology either on simulated or actual data. Both types of analyses have advantages and disadvantages, as will be described in this chapter. Nearly all studies have assumed that the basis for comparison of genomic evaluations is the parent average (PA) of genetic evaluations derived by standard mixed model methodology. The main criteria for comparison are accuracy and bias of the evaluations. In addition to analyses of simulated populations, we will describe studies that have analyzed actual dairy cattle, poultry, and swine data. All the methods described in the previous chapters have been able to outperform PA on large data sets.

Criteria for Evaluation of Estimated Genetic Values

The two most important criteria for evaluation of estimated genetic values are accuracy and bias. On simulated data accuracy can be estimated by the correlation between the estimate and the actual breeding value. The coefficient of determination, that is, the correlations squared, should equal the "reliability" of the genomic evaluation, as explained in Chapter 6 for the mixed model.

On real data the true breeding values are not known. As noted previously, genomic estimated genetic values for young animals with genotypes, but without trait records or progeny records, are compared to standard estimated breeding values (EBV) on the same animals based on progeny records produced later. In this case the squared correlation between the two estimates should equal the reliability of the genomic evaluations divided by the reliability of the daughter-based evaluations (VanRaden *et al.*, 2009). Generally statistics derived from genomic estimated breeding values (GEBV) are compared with the same statistics computed from parent average (PA). This is somewhat problematic, because if the current breeding value of the bull is based on relatively few records, or if the heritability of the trait is low, the contribution of the parents to the current genetic evaluations will still be significant. This will inflate coefficients of determination for both genomic and PA but will have a greater effect on PA (Weller *et al.*, 2014b). This problem can be somewhat alleviated if genomic and PA evaluations are compared to the bulls' current DYD. However, as noted previously, DYD are not computed for all traits.

Genomic Selection in Animals, First Edition. Joel Ira Weller.

In the previous two chapters, formulas were presented to estimate genomic reliabilities for both two- and single-step reliabilities. Thus the "realized" reliabilities can be also compared to the expected reliabilities.

Generally bias is measured by the regression of true genetic values on estimated genetic values. If evaluations are unbiased, then this regression should not be significantly different from unity. That is, the covariance of the estimated and true genetic values should equal the variance of the estimated genetic values. In other words if the difference in the estimated genetic evaluations of two animals is equal to 100 units, this should on the average be the difference between their actual breeding values. The common situation for biased evaluations is that this regression is less than unity. In this case the genetic evaluations of the top-rated individuals will tend to be higher than their actual genetic values. If bulls with daughter records are compared to bulls with genotypes, but without daughter records, the top bulls with genomic evaluations may have evaluations higher than bulls with daughter records, but these evaluations will be biased upward, and selection of individuals for breeding will not be optimal.

Methods Used to Validate Genomic Genetic Evaluations

In Chapter 14 we discussed briefly methods for validation of genomic genetic evaluations. Two basic methods have been applied in the literature. In the first method, first applied by Meuwissen *et al.* (2001), simulated data sets are derived, and genomic breeding values are computed on the simulated data. This method has the advantages that the EBV can be compared to the "true" (simulated) breeding values and that any number of simulated data sets can be generated. The main disadvantage is that it is not known how accurately the simulation algorithm actually corresponds to actual data.

The second method, first applied by VanRaden *et al.* (2009), is based on analysis of actual data. The data is divided into "training" and "validation" data sets. Generally this is accomplished by dividing the population into old and young animals, respectively. (Although the validation set comprised young animals, only animals with their own trait records or records of progeny can be used for validation.) Prediction equations that use markers, phenotypic records, and pedigree information as input and predict genomic breeding values are derived from the training data set, in which a large number of markers have been assayed on a sample of animals. Then the accuracy of the prediction equation is evaluated on the validation data set, using only genotype and pedigree information to derive genomic breeding values. The genomic breeding values of the validation animals are compared to standard breeding values for the same animals based only on pedigree and progeny records. The advantage of this method is that all the properties of actual data are intact. The disadvantages are that the number of actual data sets that can be analyzed is generally very limited and it is not possible to compare the EBV to the true breeding values.

Interbull, the international organization for validation of dairy bull genetic evaluations, validates GEBV by comparison of current deregressed EBV to GEBV derived from a truncated data set with all records of the last 4 years deleted (https://wiki.interbull.org/public/CoPAppendixVIII?action=print&rev=44). Based on the regression of current deregressed EBV on GEBV from the truncated data set, bias and increase in accuracy of GEBV as compared to PA are evaluated. Bias is estimated by the consistency of the genetic trend captured by GEBV and the consistency of variation of GEBV from the truncated data set and current EBV.

Evaluation of Two-Step Methodology Based on Simulated Dairy Cattle Data

VanRaden (2008) simulated marker and QTL inheritance for 50,000 biallelic markers and 100 biallelic QTL on 30 equal-length chromosomes. Predictions were tested with simulated genotypes for 2967 Holstein bulls and 766 Jersey bulls. The Holstein bulls included 1885 bulls born from 1995 through 1997, 290 ancestor bulls included in computing predictions, and 792 younger bulls born from 2001 through 2002 for testing predictions. The Jersey bulls included 563 older bulls to compute predictions plus 203 younger bulls to test predictions.

Reliability for Holsteins averaged 0.66 for nonlinear predictions and 0.63 for linear predictions versus 0.32 for PA for net merit of young bulls. Four methods to compute allelic frequencies for the linear model analyses were compared: the simulated frequencies, the base population frequencies (as explained in Chapter 15), simple frequencies estimated by counting alleles in genotyped bulls, and frequency of 0.5 for all markers. Differences in the reliabilities between these methods were all less than 1%. Corresponding accuracies of selection obtained as square roots of those values were 0.81, 0.79, and 0.56. Thus, linear genomic predictions had reliabilities that were 0.31 greater than the reliability for PA for the younger Holstein bulls and 0.19 greater for the younger Jersey bulls.

Numerous studies have shown that accuracies of genomic evaluations are no higher than PA, if less than 1000 bulls are included in the training population (e.g., VanRaden *et al.*, 2009; Weller *et al.*, 2014b). Since thousands of bulls with genetic evaluations based on daughter records are available only for the largest dairy cattle populations, several studies have proposed generating training populations of cows with production records and genotypes. As genotyping costs decrease, genotyping of tens of thousands of cows has become economically feasible, even for relatively small populations.

Jiménez-Montero *et al.* (2012) evaluated different female-selective genotyping strategies in populations that have a limited number of sires but a large number of cows. Populations of 40,000 cows were simulated, and 1000, 2000, and 5000 cows were selected for genotyping as training sets based on five different strategies. Genomic genetic evaluations were computed using the Bayesian Lasso algorithm (Chapter 15). The accuracy of the evaluations in the validation population using the two-tailed strategies was better than the accuracy obtained using other strategies(0.50 and 0.63 for the two-tailed selection by yield deviations strategy in low- and medium-heritability scenarios, respectively, using 1000 genotyped cows). All selection strategies resulted in biased evaluations. Increasing the number of cows genotyped to 5000 increased the accuracy by less than 10%. When 996 genotyped bulls were used as the training population, accuracies were 0.48 and 0.55 for low- and medium-heritability traits. Thus selective genotyping of cows for the training population can be more effective than genotyping a similar-sized sample of bulls.

Evaluation of Multistep Methodology Based on Actual Dairy Cattle Data

VanRaden *et al.* (2009) used genotypes for 38,416 markers and August 2003 genetic evaluations for 3576 Holstein bulls born before 1999 to predict January 2008 daughter deviations for 1759 bulls born from 1999 through 2002. Five milk yield traits, 5 fitness traits, 16 conformation traits, and net merit were analyzed. The official PA from 2003 and a 2003 PA computed from only the subset of genotyped ancestors were combined with genomic predictions using a selection index. Expected genomic reliabilities were computed from Equation (15.12). Realized genomic reliabilities (RGR) were computed as follows:

$$\mathrm{RGR} = \left(\frac{R^2}{R_{\mathrm{dau}}} \right) + \mathrm{PA\ adjustment} \tag{17.1}$$

where R^2 is the coefficient of determination between the genomic breeding value and the 2008 breeding value based on daughter records, R_{dau} is the reliability of the 2008 DYD, and the PA adjustment = reliability of the PA–the coefficient of determination of the PA and the 2008 evaluations divided by $\left(R^2/R_{\mathrm{dau}}\right)$. Since the final genomic evaluations were computed from an index that included a contribution from the PA genetic evaluation, it was necessary to account for this in calculation of RGR. The gain from genomic evaluations is the difference between the realized genomic reliability and the reliability of traditional PA.

Combined predictions were more accurate than official PA for all 27 traits. The coefficients of determination were 0.05–0.38 greater with nonlinear genomic predictions (Bayes A method of VanRaden *et al.* (2009)) compared with those from PA alone. Linear genomic predictions had coefficients of determinations similar to those from nonlinear predictions but averaged just 0.01 lower. For all traits the realized reliabilities of the genomic evaluations were about 10% lower than the theoretical reliabilities. Similarly, the realized reliabilities of the PA were generally lower than the published reliabilities. Averaged across all traits, combined genomic predictions had realized reliabilities that were 23% greater than reliabilities of PA (50 vs. 27%), and gains in information were equivalent to 11 additional daughter records.

Reduction of the number of markers by half had virtually no effect on the coefficients of determination for then genomic evaluations. Over the range of 1151–3576 predictor bulls, gains in R^2 for net merit were nearly linear with increasing numbers of predictor bulls, and gains for most other individual traits followed that same pattern. More recent results show that this trend appears to hold up to 10,000 predictor bulls. Regressions of 2008 evaluations on genomic evaluations of validation bulls were not computed in this study. Thus bias of the genomic evaluations was not estimated.

Colombani *et al.* (2013) computed GEBV for two reference populations consisting of 3940 Holstein bulls and 1172 Montbeliarde bulls using the method of VanRaden (2008) denoted GBLUP; partial least squares regression (PLS); sparse PLS (sPLS) regression, a variable selection PLS variant; and two Bayesian methods—Bayes Cπ and Bayesian least absolute shrinkage and selection operator (LASSO). Milk yield, fat content, and conception rate were analyzed. In Holsteins correlations were higher for Bayesian methods for fat content and were similar to GBLUP for milk yield and to GBLUP and PLS regression for the conception rate. Colombani *et al.* (2013) suggest that the higher correlations of the Bayesian methods for fat content are probably due to the effect of the *DGAT1* gene (Grisart *et al.*, 2002). Regression slopes of observed DYD on predicted DYD for Holsteins were less than unity for all methods in all traits but highest for standard BLUP.

Evaluation of Single-Step Methodologies Based on Actual Dairy Cattle Data

Aguilar *et al.* (2010) analyzed US Holstein data for final score used for May 2009 official evaluations. A total of 10,466,066 records were available for 6,232,548 cows. Pedigrees were available for 9,100,106 animals, and 6508 bulls were genotyped for the mid-density BeadChip, which included 54,001 markers. Genetic evaluations were calculated for 2575 young bulls with no daughter records in 2004, but with daughter records in 2009.

For PA, the coefficient of determination for the validation bulls was 0.24 and the regression was 0.76 for deregressed genetic evaluations (Chapter 15). Coefficients of determination were approximately 0.1 higher for genetic evaluations. Thus PA overestimated the genetic evaluations by 27% as compared to evaluations based on daughter records. For the multiple-step approach, the coefficient of determination increased to 40% and the regression to 0.86. The increase in coefficient of determination of 16% relative to the PA was slightly higher than the increase of 13% reported by VanRaden *et al.* (2009).

The coefficient of determination for the single-step analysis for deregressed evaluations varied from 0.37 to 0.41, and the regression varied between 0.68 and 0.79 depending on the assumed allelic frequencies for the genetic markers in computations of the **G** matrix (Equation (15.8)). The highest coefficient of determination was obtained with an assumed allelic frequency of 0.5 for all markers. Coefficients of determination were 0.38 and 0.37 if frequencies were computed from base animals and all animals, respectively. As explained in Chapter 15, "base animals" are animals with genotypes, but without ancestors with genotypes. Regressions were highest, 0.79, with allelic frequencies computed by the method of Gianola *et al.* (2009), as compared to a regression of 0.76 if allele frequencies of 0.5 were assumed.

Coefficients of determination were highest assuming that the **G** matrix accounted for all genetic variance. That is, $\lambda=1$ in Equation (16.10). Regressions increased as λ decreased to 0.5, but coefficients of determination also decreased slightly. With $\lambda=0.6$ the coefficient of determination was 0.4 and the regression was 0.90. Thus a substantial increase in the regression was obtained with only a minor reduction in the coefficient of determination. With this value for λ results for the single-step methodology have a coefficient of determination equal to the multistep methodology, but a regression of 0.9, as compared to 0.86 for the two-step methodology. Of course it must be noted that these results are based on a single trait for a single data set, with λ set empirically at the optimal value.

Lourenco *et al.* (2014a) evaluated methods for derivation of GEBV for an Israeli Holstein dairy population of 713,686 cows and 1305 progeny-tested bulls with genotypes. Inclusion of genotypes of 343 elite cows in an evaluation method that considers pedigree, phenotypes, and genotypes simultaneously also was evaluated. For each production trait, a multitrait animal model was used to compute traditional genetic evaluations for parities 1 through 3 as separate traits. On average, R^2 was lowest for PA followed by the method of VanRaden (2008), Bayes C, single-step methodology with all markers weighted equally, and single-step methodology with differential marker weights (Wang *et al.*, 2012).

Evaluation of Single- and Multistep Methodologies Based on Actual Poultry Data

Unlike dairy cattle, in poultry it is not possible to generate a validation population, consisting of a large number of sires genotyped each with many progeny. Instead the validation population consisted of animals with single records for the traits analyzed. Thus accuracy of the GEBV, the correlation between the estimated and actual breeding values, were computed as

$$r(\hat{u}, u) = \frac{r(\hat{u}, u+e)}{h} \tag{17.2}$$

where $\hat{u}$ is the genetic evaluations of validation animals based on genotypes, but without records, $u+e$ is the phenotypic records of these animals corrected for fixed effects, and h is the square root of heritability.

Data of broiler chickens for two pure lines for three generations were analyzed (Simeone *et al.*, 2011). The complete population included 183,784 and 164,246 broilers for the two lines, of which 3284 and 3098 broilers were genotyped for 57,636 SNPs. The validation population consisted of all third-generation birds of which 799 individuals were genotyped from each line. The average accuracies of the validation population with standard mixed model evaluations for body weight at 6 weeks, breast meat, and leg score were 0.46, 0.30, and less than 0. Accuracies with single-step genomic evaluations were 0.60, 0.34, and 0.06, respectively, and 0.60, 0.36, and 0.09 for two-step Bayes A methodology. On the low heritability trait of legs score the single-step methodology was 50% more accurate than the standard mixed model considering all animals of the third generation. A similar gain has not been demonstrated for dairy cattle, but it should be noted that PA evaluations are probably less reliable in poultry due to the lower number of progeny per sire.

Evaluation of Single- and Multistep Methodologies Based on Actual Swine Data

A swine data set consisting of 3534 animals from a single nucleus pig line with genotypes from the Illumina PorcineSNP60 chip and a pedigree including parents and grandparents of the genotyped animals for a total of 6473 animals was analyzed (Cleveland *et al.*, 2012). Five traits with heritabilities ranging from 0.07 to 0.62 were recorded, but the number of animals with records varied among the traits analyzed. Genomic breeding values were calculated using Bayes B with phenotypes and with deregressed breeding values and using a single-step genomic BLUP approach with information from both genotyped and ungenotyped animals.

In each analysis one-sixth of the genotyped animals were randomly assigned to the validation set and the remaining genotyped animals to the training set. Each training set therefore consisted of 2945 genotyped animals. For the animals in the validation set only genotype information and pedigree were used to compute genomic genetic evaluations. Accuracy of the genomic breeding values for the validation animals was estimated as the correlation between genomic and high accuracy mixed model breeding values based on progeny records. Only 75 of the animals in the validation set with the highest reliabilities were used to estimate accuracy of the genomic breeding values, but no correction was made for the fact that the mixed model breeding values do not correspond completely to the true breeding values.

Genomic evaluations were also computed with the 509 youngest animals as the validation set and all older animals in the training set. Due to the low reliabilities of the young animals, accuracies were estimated using the 30 validation animals with the highest mixed model accuracies for each trait, but only the three traits with the highest heritabilities were analyzed.

The genomic breeding value accuracy increased with increased trait heritability and with increased relationship between training and validation. In nearly all cases, Bayes B using deregressed breeding values outperformed the other approaches, but the single-step evaluation performed only slightly worse. Accuracies for the Bayes B methodology with the young boars included in the validation set ranged from 0.5 to 0.7 for the three traits with heritabilities of 0.38–0.62.

Evaluation of GEBV for Plants Based on Actual Data

Nearly all studies that have analyzed plant data have used methods described here as multistep, although for plants generally all individuals that were phenotyped were also genotyped. Thus, there is no justification for single-step methodologies. In all cases the number of markers and individuals genotyped was much lower than in the animal studies (reviewed by Nakaya and Isobe, 2012). Thus a number of studies have been able to apply standard BLUP to estimate marker effects. Other estimation methods have been used including Bayesian LASSO and ridge regression.

The ranges of accuracies in empirical studies were higher in plant studies than animal studies, despite the relative low numbers of markers and individual genotypes. In a few cases accuracies about 0.9 were obtained. Nakaya and Isobe (2012) suggest that this might be due to the lower genetic diversity caused by a small number of parental lines and a greater bottleneck in the breeding materials.

Summary

From the studies described in this chapter and additional studies not presented in detail, several important conclusions can be drawn. First, the major factor affecting the accuracy of genomic evaluations is the number of animals with genotypes used to derive genomic evaluations. Derivation of genomic evaluations with significantly greater accuracy than PA requires a training population of at least several thousand genotyped individuals. The number of markers included in the analysis over the range tested has only a very minor effect on accuracy. Thus very little gain can be expected by the introduction of high-density DNA chips including up to 800,000 markers. Some bias is nearly always observed. Thus the young individuals with the highest genomic evaluations are inflated. However, this is generally the case also with PA. Differences in accuracy between single- and multistep methodologies were minimal, but single-step methodologies might have an advantage in bias, and for genetic evaluation of young animals that were not genotyped.

18 By-Products of Genomic Analysis: Pedigree Validation and Determination

Introduction

The availability of high-density SNP chips and more recently next-generation whole genome sequencing has additional by-products, in addition to genomic genetic evaluations. In this chapter we will review the application of genetic markers to pedigree validation and determination. Until the advent of microsatellites in the 1990s, parentage verification could only be performed based on blood groups, and costs were nearly prohibitive, except for specific animals of particular interest. Now with routine genotyping of hundreds of thousands of individuals for low- and medium-density SNP chips, parentage verification is becoming a routine procedure for a large fraction of the commercial population. Pedigree mistakes lower rates of genetic gain. Confirmation and correction of pedigrees can be considered a by-product of genomic evaluation obtained at virtually no extra cost. As genotyping costs decrease, the gains obtained directly from pedigree correction may cover genotyping costs without considering gains due to genomic selection. This is clearly a "win–win" situation. We will first consider the effects of incorrect pedigree information on breeding programs. We will then review the history of pedigree validation prior to SNP chips and explain in detail the current state of the art for both pedigree validation and determination based on low- and medium-density SNP chips.

The Effects of Incorrect Parentage Identification on Breeding Programs

Van Vleck (1970a, 1970b) demonstrated for sire models that incorrect identification of sires can bias estimates of heritability and genetic evaluations and reduce genetic progress due to selection. Israel and Weller (2000) estimated the effect of pedigree errors on estimated breeding value and genetic gain derived by a single trait animal model for a sex-limited trait with a heritability of 0.25. These values correspond approximately to the three major milk production traits. Ten populations of 100,000 milking cows were simulated with correct paternity identification for all animals, and 10 populations were simulated with 10% incorrect paternal identification. The initial populations consisted of 100,000 unrelated individuals, and simulations were continued for 20 years. The BLUP genetic evaluations were computed every year by an animal model analysis for each complete population. Estimated breeding values for the populations with 10% incorrect paternity were biased, especially in the later generations. Genetic gains were 4.3% higher with correct paternity identification.

Genomic Selection in Animals, First Edition. Joel Ira Weller.

Principles of Parentage Verification and Identification with Genetic Markers

Baring mutations and genotyping mistakes, a progeny and each parent must have one allele in common at each genetic marker. Weber and Wong (1993) found an average mutation rate for 28 microsatellites on human chromosome 19 of 1.2×10^{-3} per locus per gamete per generation. This rate may have been inflated by somatic as opposed to germ-line events. For all genetic markers considered, rates of genotype mistakes are always much higher than mutation rates.

Parentage is considered "confirmed" if the null hypothesis of correspondence by chance between the genotypes of the progeny and the putative parent can be rejected with sufficiently high probability. This probability is computed based on the allelic probabilities in the general population. For example, considering a single marker, if the progeny is heterozygous for alleles i and j, the probability that another random individual in the population could not be rejected as a parent will be $1-(1-p_i-p_j)^2$, where p_i and p_j are the population probabilities for alleles i and j. If the progeny is homozygous for allele i, then this expression reduces to $1-(1-p_i)^2$. If several markers are considered, then the overall probability for nonexclusion for individual i, Nx_i, will be

$$Nx_i = \prod_{j=1}^{J}\left[1-\left(1-p_k\right)^2\right] \quad (18.1)$$

where p_k = sum of the frequencies of the two alleles of individual i for marker j and $\Pi[*]$ represents the product of [*] from 1 through J, where J is the number of markers for which individual i and its putative sire were genotyped.

The probability to reject paternity (exclusion probability) for an erroneously recorded parent will increase with an increase in the "polymorphism information content" (PIC) of the marker, defined as the probability of obtaining a progeny for which allele origin of a single parent can be determined from a random mating (Botstein *et al.*, 1980). If both parents and the progeny are genotyped, then PIC is computed as follows:

$$\mathrm{PIC}_2 = 1-\sum_{i=1}^{\mathrm{Na}} p_i^2 - \sum_{i=1}^{\mathrm{Na}-1}\sum_{j=i+1}^{\mathrm{Na}} p_i^2 p_j^2 \quad (18.2)$$

where PIC_2 = PIC with both parents genotyped, Na is the number of marker alleles segregating in the population, and p_i and p_j are the population frequency of the two progeny alleles. It can be easily seen that if only two alleles are segregating in the population, PIC_2 will be maximum with $p_i = p_j = 0.5$.

If only a single parent and the progeny are genotyped, then PIC is computed as follows:

$$\mathrm{PIC}_1 = 1-\sum_{i=1}^{\mathrm{Na}} p_i^2 - \sum_{i=1}^{\mathrm{Na}-1}\sum_{j=i+1}^{\mathrm{Na}} p_i p_j \left(p_i + p_j\right) \quad (18.3)$$

where PIC_1 = PIC with a single parent genotyped. For a given number of alleles, PIC will be maximum if the frequencies of all alleles are equal, that is, $p_1 = p_2 = \cdots = p_{\mathrm{Na}} = 1/\mathrm{Na}$. For all numbers of alleles and all allelic frequencies, $\mathrm{PIC}_2 > \mathrm{PIC}_1$.

The probability that allele origin of the progeny can be determined for both parents is termed the "proportion of fully informative matings" (PFIM) (Haseman and Elston, 1972). PFIM is computed as follows (Gotz and Ollivier, 1992):

$$\text{PFIM} = \sum_{i=1}^{\text{Na}}\sum_{j=1}^{\text{Na}-1} p_i^2 p_j^2 \left[\left(\sum_{k=1}^{\text{Na}-1}\sum_{l=k+1}^{\text{Na}} 2p_k p_k\right) - 2p_i p_j\right] \tag{18.4}$$

where p_k is the combined probability of all alleles other than i and j in the population, and the other terms are as defined previously. With all alleles at equal frequency, this equation reduces to

$$\text{PFIM} = \frac{(\text{Na}-1)(\text{Na}-2)(\text{Na}+1)}{\text{Na}^3} \tag{18.5}$$

Paternity Validation Prior to High-Density SNP Chips

Beginning in the 1940s parentage could be determined in cattle using blood group markers(Stormont *et al.*, 1951), but the procedure was very expensive and was generally only applied to validate the parentage of AI sires (Stormont, 1967). Because of the limited number of markers and low informativeness of these markers, it was generally necessary to have the genotypes of both parents to obtain reasonable power to reject the hypothesis that correspondence was obtained by chance. Various studies found that the proportion of errors in sire identification varied from a few percent to as much as 22% (Christensen *et al.*, 1982; Geldermann *et al.*, 1986; Bovenhuis and van Arendonk, 1991).

With the advent of DNA microsatellites, it became possible to verify parentage of large numbers of animals as a by-product of daughter and granddaughter designs conducted during the 1990s and 2000s. With respect to parentage verification, microsatellites had the advantage that they were generally multiallelic (Glowatzki-Mullis *et al.*, 1995). Thus even if the progeny was heterozygous for a specific marker, there was generally a significant probability that the progeny would have no common allele for an incorrectly recorded parent. The disadvantage of microsatellites was that genotyping error rates were generally high, in the range of 5–10%. Thus, at least two discrepancies between the progeny and the putative parent were generally required to reject parentage (Weller *et al.*, 2004).

Only a very few studies actually reported on the prevalence of parentage mistakes in commercial populations. Weller *et al.* (2004) analyzed a total of 6040 Israeli Holstein cows genotyped for 104 microsatellites from 181 herds listed as progeny of 11 sires were. The frequency of rejected paternity was 11.7%. The effects of recorded sire, birth year, geographical region, herd, and inseminator on the frequency of paternity rejection were analyzed with linear and nonlinear models. Only the effects of inseminator and recorded sire were significant in all models tested that included these effects. The main causes of incorrect paternity recording appear to be inseminator recording mistakes and possibly mistakes with respect to semen labeling at the AI institutes. Incorrect paternity recording due to multiple inseminations by different sires could explain at most 20% of the paternity mistakes.

The International Society for Animal Genetics (ISAG) proposed specific panels of microsatellites to validate parentage for all the major commercial animal species (e.g., Bredbacka and Koskinen, 1999). Although microsatellites could efficiently validate parentage, due to the relatively high genotyping error rate, they could not be used to determine the actual parent from a list of candidates, unless the list was very limited.

Paternity Validation and Determination with SNP Chips

SNPs are nearly always biallelic. Thus exclusion of a putative parent is obtained only when one individual is homozygous for one allele and the other individual is homozygous for the other allele. The advantages of SNPs relative to microsatellites are that genotyping error rates are much lower and the number of SNPs included on medium-density SNP chips is much greater than the number of microsatellites generally available for analysis. A standard subset of approximately 100 SNP (Heaton *et al.*, 2002) is included in almost all genotyping chips and has been accepted for international parentage confirmation (International Committee for Animal Recording). By analysis of 40,874 valid SNPs, Wiggans *et al.* (2009) found that the mean number of conflicts was 2.3 when pedigree was correct and 2411 when it was incorrect. Between 4 and 14% of US dairy cattle genotyped with the Illumina Bovine3K chip had incorrectly reported sires (Wiggans *et al.*, 2012a).

Of the 576 Israeli Holstein bulls genotyped by the BovineSNP50 BeadChip (Weller *et al.*, 2010), there were 204 bulls for which the father was also genotyped. The results of 38,828 valid SNPs were used to validate paternity, determine the genotyping error rates, and determine criteria enabling deletion of defective SNPs from further analysis. Based on the criterion of greater than 2% conflicts between the genotype of the putative sire and son, paternity was rejected for seven bulls (3.5%). The remaining bulls had fewer conflicts by one or two orders of magnitude. Excluding these seven bulls, all other discrepancies between sire and son genotypes are assumed to be caused by genotyping mistakes. The frequency of discrepancies was greater than 0.07 for 9 SNPs and greater than 0.025 for 81 SNPs. The overall frequency of discrepancies was reduced from 0.00017 to 0.00010 after deletion of these 81 SNPs, and the total expected fraction of genotyping errors was estimated to be 0.05%, as compared to genotyping error rates of 5–10% with microsatellites.

Validation of More Distant Relationships

As noted previously, baring mutations and genotyping mistakes, a parent and a progeny must have one allele in common for each genetic marker. However, this is not the case for more distant relationships, such as grandprogeny–grandparent or half-sibs. However, related individuals should still have more alleles in common as compared to unrelated individuals.

Seroussi *et al.* (2013) proposed a method to validate maternity based on the genotypes of their sons and fathers. The method was applied to 789 Israeli AI bulls genotyped for the BovineSNP50 BeadChip. They visualized the pairwise identity-by-state distances calculated using PLINK software in three dimensions (Purcell *et al.*, 2007). Each of the 310,866 possible pairs of individuals (789*788/2) was represented by the three dimensional coordinates that correspond to the frequency of the three possible states of their SNPs' alleles: no match ($f0$), single match ($f1$), or double match ($f2$). Results were reduced to two dimensions using the transformations: $x' = 0.7071(1 + f1 - f2)$ and $y' = 1.2247(f0)$. Bull-by-bull pairs were grouped according to their level of kinship, and canonical scores were calculated using discriminant analysis and the x' and y' features. Of the 474 pairs of recorded maternal grandsire–grandson with both individuals genotyped, the probability for 28 pairs to belong to the grandsire–grandson level of kinship was less than 5%, which postulates an error rate of around 3% per generation in pedigree determination.

VanRaden *et al.* (2013a) used three methods to validate and determine maternal grandsire and maternal great-grandsires. In the first method, only the progeny and the putative maternal grandsire genotypes were used. The number of discrepancies was counted, with "discrepancy" defined as given previously. That is one individual homozygous for one allele and the other individual homozygous for the other allele, even though there is only a 50% chance that the grandprogeny received either allele from its grandsire. In the second method, the genotype of the sire was also included in the analysis. This method also counts conflicts using heterozygous loci if the sire is homozygous, because the allele contributed by the dam is then known. The first and second methods were considered to have produced a confirmation if the potential maternal grandsire with the lowest fraction of discrepancies was the reported pedigree maternal grandsire.

The third method imputes genotypes for all markers with missing genotypes and counts the haplotypes in common instead of individual SNP conflicts. (Methods for imputation of missing genotypes will be considered in detail in the next chapter.) The paternal haplotype is removed from the animal's genotype (similar to the second method) to determine the maternal contribution. A match is declared if the maternal haplotype is the same as either of the two maternal grandsire haplotypes. Again there is only a 50% chance the grandprogeny received either of the grandsire's haplotypes. In this method the correct maternal grandsire was considered confirmed if the bull with the highest fraction of matches was the recorded maternal grandsire.

The three methods were applied to 12,152 Holstein, 2265 Jersey, and 1605 Brown Swiss potential grandsires. The correct maternal grandsire was selected with frequencies of 61, 60, and 65%, respectively, with the first method; 95, 91, and 94% with the second method; and 97, 95, and 97% with the third method.

Pedigree Reconstruction with High-Density Genetic Markers

With the advent of high-density SNP chips, it became possible to reconstruct complete pedigrees if sufficient numbers of individuals are genotyped (Anderson and Garza, 2006; Hill *et al.,* 2008; Gorbach *et al.,* 2010). In natural populations, no pedigrees were recorded historically and reconstruction of the pedigree information from DNA may be a very cost-effective breeding option (Pemberton, 2008; El-Kassaby and Lstibůrek, 2009).

In Equation (15.8) we presented the formula for construction of the genomic relationship matrix. Seroussi *et al.* (2013) calculated the genomic relationship for 789 bulls genotyped for the BovineSNP50 BeadChip. The average score observed for the entire third group of kinship (half-sibs and grandparent–grandprogeny pairs) was 0.25, which matches the theoretical value. The number of misclassified pairs using this analysis, as compared to the pairwise identity-by-state distances calculated using PLINK software described previously, was reduced by 3.4 (7.8%).

Summary

Methods to verify recorded genetic relationships began in the 1940s with blood group markers but only became really cost effective with the advent of microsatellites in the 1990s. Estimates of the frequency of incorrect parentage recording range from a few percent to over 20%, with a mean near 10%. Although SNPs are nearly always diallelic, the lower error rates, and the much greater number of markers genotyped, resulted in much more efficient pedigree validation, compared to microsatellites. With the development of mid-density SNP chips with approximately 50,000 markers, determination of actual pedigree relationships, even removed two generations, became possible. Since pedigree errors reduce rates of genetic gain, the ability to correct pedigree mistakes is a virtually zero cost dividend of genomic selection programs.

19 Imputation of Missing Genotypes: Methodologies, Accuracies, and Effects on Genomic Evaluations

Introduction

As noted in the previous chapters, low-, medium-, and high-density BeadChips are now available for all the important agricultural species (e.g., http://support.illumina.com/array/kits.ilmn). Typically low-density chips contain less than 10,000 SNPs, while medium-density chips contain over 50,000 SNPs, and high-density chips contain close to 800,000 SNPs. Current costs per individual genotyped are in the range of \$40, \$90, and \$200 for low-, medium-, and high-density chips, respectively. (By the time this text reaches publication, these numbers will probably be lower.) Hundreds of thousands of cows have already been genotyped in the United States for low-density chips, and similar situations exist in other farm animals. Various studies have proposed that the "missing" genotypes in low-density chips can be inferred based on the prevalence of haplotypes in the population. That is, the haplotype for a specific chromosomal location of the individual genotyped for the low-density chip is compared to haplotypes prevalent in the population of individuals genotyped for the medium- or high-density chip. By matching the low-density haplotype to the corresponding haplotype for individuals genotyped for the medium- or high-density chip, the SNPs missing in the low-density chip can then be deduced, considering also pedigree and prevalence of haplotypes in the population, as illustrated in Figure 19.1.

This procedure has been termed "imputation." Of course the same method can also be used to deduce specific missing genotypes on medium-density chips and to correct "erroneous" genotypes—that is, genotypes that conflict with Mendelian rules of inheritance, as explained in the previous chapter. In the first sections we will consider the different algorithms that have been proposed for imputation and explain why different strategies have been proposed for humans and farm animals. In the following sections we will compare these methods on actual cattle data, based on accuracy and speed, and in the final section we will consider how imputation affects genomic evaluation of farm animals.

Determination of Haplotypes for Imputation

As explained in the introduction all methods of imputation require determination of haplotypes. There are two general methods to determine haplotypes from genotypes: "statistical"- and "pedigree"-based methods. In the first method, haplotypes of most individuals are determined based

Genomic Selection in Animals, First Edition. Joel Ira Weller.

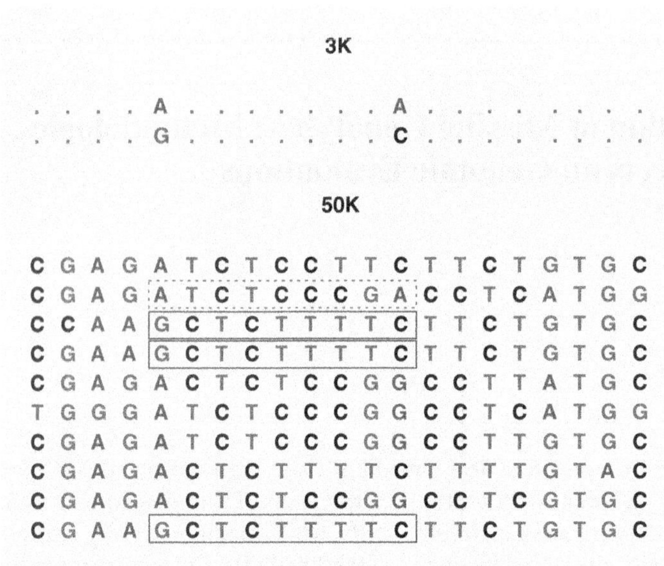

Figure 19.1 Illustration of imputation based on common haplotypes for a specific chromosomal segment. Only two markers are assumed to be genotyped on the low-density (3K) chip and 21 markers on the medium-density (50K) chip. The two haplotypes of a single individual genotyped for the low-density chip are shown in the top part of the figure, and the pair of haplotypes for five individuals genotyped for the medium-density chip are illustrated in the lower part of the figure. Only the second haplotype listed for the medium-density chip corresponds to the first haplotype of the individual genotyped for the low-density chip and is indicated with a dotted line box. Three haplotypes of the medium-density chip correspond to the second haplotype for the individual genotyped for the low-density chip, and these are indicated with solid line boxes. All three have the same haplotype. Thus by imputation we can assume that the missing genotypes of the individual genotyped for the low-density chip correspond to the second and third rows in the lower part of the figure.

on the frequencies of specific haplotypes in the population. In the second method, haplotypes of parents first are determined based on the genotypes of their progeny, which are then used to iteratively determine the genotypes of the progeny (e.g., Baruch *et al.,* 2006). The first method is more computing intensive and is generally applied to human populations, in which the numbers of related individuals with genotypes are usually very limited.

Imputation in Humans versus Imputation in Farm Animals

The first algorithms for imputation were developed based on human populations. Analysis of human pedigrees is generally based on nuclear families of parents and a small number of full sibs, while farm animals generally have large half-sib families. Furthermore, in humans the individual's genotypes will be selected based on the inheritance of a specific syndrome in particular families. In farm animals the current situation is that nearly all sires used for artificial insemination are genotyped for medium- or high-density BeadChips, while females, daughters of these sires, are genotyped for low-density chips. This facilitates determination of haplotypes in the sires by pedigree-based methods, and this information can then be used to determine the haplotypes of progeny.

Algorithms Proposed for Imputation in Human and Animal Populations

For human populations several imputation methods based on various statistical models, such as the haplotype clustering algorithm (Scheet and Stephens, 2006), the hidden Markov model (Browning and Browning, 2007), the expectation maximization (EM) algorithm (Qin *et al.*, 2002; Scheet and Stephens, 2006), and the Markov chain model (Li *et al.*, 2010), have been proposed. Findhap, unlike the previous algorithms, which were developed to deal with human data, is a pedigree-based algorithm developed for farm animals, especially dairy cattle (VanRaden, 2011). The principles of fastPHASE, BEAGLE, IMPUTE, and Findhap will now be discussed in detail.

fastPHASE uses a localized haplotype clustering algorithm (Scheet and Stephens, 2006). It assumes that haplotypes of individuals in the population tend to cluster into groups of closely related or similar haplotypes within a short region of a chromosome. This method allows memberships of clusters to vary along the chromosome based on a hidden Markov model. Missing genotypes are sampled on the basis of allele frequencies estimated from reference haplotypes, and then an EM algorithm is used to estimate parameter values to infer missing genotypes. The computing time of the fastPHASE algorithm increases linearly with the number of ungenotyped individuals and the number of haplotype clusters (Weigel *et al.*, 2010).

BEAGLE is also a localized haplotype clustering-based algorithm (Browning and Browning, 2007). First, it gathers haplotype clusters at each marker and defines a hidden Markov model to find the most likely haplotype pairs based on the known genotypes of each individual. The most likely genotype at the missing genotype loci can then be deduced from final haplotype pairs. BEAGLE, unlike fastPHASE which estimates parameters for cluster configuration using an EM algorithm, uses empirical frequencies. Also, unlike fastPHASE, which relies on a fixed number of haplotype clusters to form underlying hidden states in the Markov chain, BEAGLE allows the cluster number to dynamically change to better fit localized linkage disequilibrium patterns (Pei *et al.*, 2008).

The program "IMPUTE" is also a hidden Markov model-based algorithm. Rather than simultaneously estimating missing genotypes and integrating over the unknown phase of SNP that are present in both the reference panels and the study sample, this algorithm estimates haplotypes at SNP that are present in both populations and then imputes genotypes in the study sample, assuming that these haplotype guesses are correct. Uncertainty about phasing is taken into account by iterating these steps in a Markov chain Monte Carlo framework. Thus, unlike many competing algorithms for which phasing accuracy does not depend on the size of the study sample, IMPUTE gains accuracy by using information from both the reference panels and study sample during the phasing step. The computational feasibility of IMPUTE is enhanced by using only a subset of haplotypes at each iteration to build the conditional distribution of haplotypes of observed SNP for an animal in the study sample, given the animal's genotype, the haplotypes of other animals in the study sample, and the haplotypes of animals in the reference panel. Rather than sampling these "conditioning states" randomly, this algorithm selects sets of haplotypes that are closest to the animal in question, based on the Hamming distance (i.e., the minimum number of substitutions required to change one haplotype into the other) between the current-guess haplotype for this animal and for other animals in the population.

Findhap (VanRaden *et al.*, 2011a) is designed to integrate the population with pedigree haplotyping. It is the only program designed specifically for farm animal populations. The steps in the algorithm are as follows:

1. Each chromosome was divided into segments with three progressively shorter lengths, long lengths to lock in identity by descent, and short lengths to fill in missing calls.
2. The first genotype was entered into the haplotype list as if it was a haplotype.

3. Any subsequent genotypes that shared a haplotype were then used to split the previous genotypes into haplotypes.
4. As each genotype was compared to the list, a match was declared if no homozygous loci conflicted with the stored haplotype.
5. Any remaining unknown alleles in that haplotype were imputed from homozygous alleles.
6. The individual's second haplotype was obtained by subtracting its first haplotype from its genotype, and the second haplotype was checked against remaining haplotypes in the list. If no match was found, the new genotype (or haplotype) was added to the end of the list. Unknown alleles in the genotype were stored as unknown alleles in the haplotype.
7. The list of currently known haplotypes was sorted from most to least frequent as haplotypes were found for efficiency so that more probable haplotypes were preferred.

In continued iterations, earlier created genotypes are matched again using haplotypes that occurred later. The first two iterations mainly focus on determination of haplotypes in the population. Only the highest-density genotypes are used in the first iteration, and then all genotypes are used in the second iteration. After haplotyping, haplotypes are matched by using both pedigree and population in the following two iterations. Known haplotypes of genotyped parents were checked first, and if either of the individual's haplotypes was not found with this quick check, then checking restarted from the top of the sorted list.

Comparisons of Accuracy and Speed of Imputation Methods

The accuracy of imputing missing genotypes using different haplotype reconstruction methods has been mostly compared using real data in humans (Marchini *et al.*, 2007; Pei *et al.*, 2008; Nothnagel *et al.*, 2009; Shriner *et al.*, 2010; Weigel *et al.*, 2010). Several studies have also been conducted for animal populations, based on simulated data and real dairy cattle data. Two studies based on real data and one study based on simulated data will be considered in this section. In the first two studies all animals were genotyped for medium-density 50K BeadChip. A fraction of the genotypes of a sample of animals are "masked," that is, assumed to be unknown, and the imputation program is applied to determine the masked genotypes. Accuracy of imputation is then computed as the fraction of correctly determined genotypes among the masked genotypes.

Weigel *et al.* (2010) evaluated the accuracy of fastPHASE 1.2 and IMPUTE 2.0 imputation in Jersey cattle, using reference panels comprising 2542 animals with 43,385 SNP genotypes and study samples of 604 animals for which genotypes were assumed known for 1, 2, 5, 10, 20, 40, or 80% of loci. The mean proportion of genotypes imputed correctly ranged from 0.659 to 0.801 when 1–2% of genotypes were available in the study samples, from 0.733 to 0.964 when 5–20% of genotypes were available, and from 0.896 to 0.995 when 40–80% of genotypes were available. When the proportion of masked genotypes was large, such as 98 or 99%, IMPUTE 2.0 was slightly more accurate, with gains in accuracy relative to fastPHASE 1.2 ranging from approximately 0.02 to 0.07. However, IMPUTE 2.0 was significantly more accurate for scenarios in which 90 or 95% of genotypes were masked in the study sample.

Weng *et al.* (2013) compared the efficiency of fastPHASE, BEAGLE, and Findhap using Chinese Holstein cattle genotyped for the Illumina BovineSNP50 genotypes. A total of 2108 cattle were randomly divided into a reference population and a test population to evaluate the influence of the reference population size. Three bovine chromosomes, 1, 16, and 28, were used to represent large, medium, and small chromosome size, respectively. They randomly masked 20, 40, 80, and

95% of the genotypes on each chromosome in the test population, but did not mask genotypes in the reference population. This corresponds to the situation in commercial populations in which bulls are generally genotyped for the medium-density chip and cows are genotyped for the low-density chip.

The three methods showed comparable accuracy when the proportion of masked SNPs was low. However, the difference became larger when more SNPs were masked. BEAGLE performed the best and was most robust with imputation accuracies of greater than 90% in almost all situations. Differences in accuracy between BEAGLE and Findhap were less than 5% in all cases. fastPHASE was affected by the proportion of masked SNPs, especially when the masked SNP rate was high. Findhap ran the fastest, whereas its accuracies were lower than those of BEAGLE, but higher than those of fastPHASE. Computing times for Findhap ranged from 20 to 50 seconds per chromosome, while computing times for BEAGLE were 1.6–18.5 hours per chromosome if at least 80% of the markers were masked, and even longer for fastPHASE. Thus considering the huge difference in computing time, and the relatively small differences in accuracy, Findhap is clearly the current method of choice for dairy cattle.

VanRaden *et al.* (2013b) found that imputation to high-density chips with Findhap gave 99.3% correct genotypes from medium density, 96.1% from 6K, and 93.7% from 3K, respectively, on a simulated chromosome. Thus imputation from medium- to high-density chips is nearly completely accurate.

Effect of Imputation on Genomic Genetic Evaluations

There are only a few studies that attempted to estimate the effect of imputation on genomic evaluations. Chen *et al.* (2011) studied the effect of imputation from the Illumina Bovine3K BeadChip to the Illumina Bovine50K BeadChip. (It should be noted though that the 3K BeadChip has been replaced with the 9K BeadChip, so results are no longer relevant on a commercial level.) They analyzed German Holstein bulls, EuroGenomics Holstein bulls, and all genotyped animals of German Holstein breed using three imputation programs: Findhap, BEAGLE, and DAGPHASE (Druet and Georges, 2010, version 2.3). A total of 1369 youngest German Holstein bulls, born between September 2003 and December 2004, were chosen as validation animals. As found by Weng *et al.* (2013), Findhap was much faster than BEAGLE and DAGPHASE. Allele error rate for the EuroGenomics bull data set was 3.3% for Findhap, 2.7% for DAGPHASE, and 1.6% for Beagle, respectively.

Genomic evaluations based on imputed data were computed only for BEAGLE and Findhap. Phenotypic data from April 2010 Interbull evaluation were used to assess the loss in accuracy of genomic prediction using the imputed 54K genotypes of EuroGenomics data set. Equal regression coefficients were obtained with the imputed 54K genotypes compared to the actual genotype, indicating that the genomic evaluations derived from imputed genotypes were no more biased than the complete genotypes. On average, reliability of GEBV dropped by 6.5% for Findhap and 2.6% for BEAGLE, respectively. Again BEAGLE outperformed Findhap with respect to accuracy, but computing time was 700-fold greater for BEAGLE.

VanRaden *et al.* (2013b) compared genomic genetic evaluations on animals genotyped with the high-density bovine BeadChip to imputed genotypes derived from masking those markers not present on the medium-density chip. They used imputed genotypes and August 2008 phenotypes to predict deregressed evaluations of US bulls proven after August 2008. Although 777,962 markers are included on the BovineHD BeadChip, only 311,725 markers were used for genomic evaluations, due chiefly to redundancies of markers. For 28 traits tested, the estimated genomic reliability averaged 61.1% when using 311,725 markers, 60.7% when using 45,187 markers (i.e., the markers present on the medium-density chip), and 29.6% for parent averages. Increasing the number of markers from 45,187 to 311,725 gave only a 0.4% point gain in average reliability of genomic predictions. These results correspond to the previous results of VanRaden *et al.* (2009) presented in Chapter 15 that reducing the number of markers relative to the medium-density BeadChip had a very minor effect on the accuracy of genomic evaluations.

Summary

Over the last 5 years, imputation has become a standard technique for increasing the accuracy of genomic evaluations derived for individuals genotyped for low-density SNP chips. Imputation has been found to be quite accurate for most situations of practical importance, and very efficient algorithms have been developed for animal populations, in which the numbers of paternal half-sib are generally very large. Thus computing costs for imputation are currently insignificant with respect to the additional costs of genotyping large numbers of animals for medium-density chips. This of course may change in the near future as genotyping costs continue to decline. Although imputation of medium-density genotypes to high-density genotypes is very accurate, it results in at best only a very minor gain in the accuracy of genomic evaluations.

20 Detection and Validation of Quantitative Trait Nucleotides

Introduction

As described in the previous chapters, all the methods currently used for genomic evaluation are based on population-wide linkage disequilibrium (LD) between the genetic marker and the actual polymorphisms responsible for this variation, the quantitative trait nucleotides (QTN).

A huge number of genome-wide association studies (GWAS) have been conducted in many species based on medium- or high-density SNP chips (reviewed by Gondro *et al.* (2013)). In Chapter 9 we considered the question of the "missing heritability." Human height has a heritability of approximately 0.9. Yet, despite the huge sample sizes and huge numbers of markers analyzed in GWAS for this trait, the sum of significant effects detected accounted for only 5% of the variance for height. Similar results were found for autism and schizophrenia, even though both diseases also have very high heritabilities. Maher (2008) gave several explanations for these disappointing results. Suggested explanations include that heritability estimates may be inflated, that a large fraction of the variance is due to copy number variations which are not detected in GWAS, the existence of gene-by-gene or gene-by-environment interactions, the common disease–rare variant hypothesis, and the possibility that inherited epigenetic factors cause resemblance between relatives.

However, it appears that the simplest explanation is the most correct. Yang *et al.* (2010) showed that 45% of variance can be explained by considering all SNPs simultaneously. Thus, most of the heritability is not "missing" but has not previously been detected because the individual effects are too small to pass stringent significance tests.

Considering the huge number of candidate polymorphisms for a QTN, the fact that population-wide linkage disequilibrium (LD) can extend over several centimorgans (Farnir *et al.*, 2000), and the biological limitations of research on farm animals, the question arises as to what constitutes proof that a QTN has in fact been determined.

As noted in Chapter 9, Ron and Weller (2007) presented a schematic strategy for QTN determination and verification in farm animals. In the current chapter we will first review methods to detect segregating QTL in animal populations based on GWAS. Strategies for QTL determination and validation will be considered in detail. We will also consider the question of what gains can be expected by QTN determination and finally review the current state of the art.

Genomic Selection in Animals, First Edition. Joel Ira Weller.

GWAS for Economic Traits in Commercial Animals

In Chapter 10 we discussed the multiple comparison problem in relation to GWAS. Since thousands of tests are generally performed, the usual significance thresholds of 0.05 or 0.001 are meaningless. In order to obtain genome-wide significance, much more stringent significant thresholds must be applied, as explained in Chapter 10. This is one of the reasons that so few significant QTL are generally detected in GWAS. Despite the generally disappointing results for GWAS to detect QTL in human populations, results have generally been somewhat more positive for animal populations, especially dairy cattle. GWAS for dairy cattle have generally been based on analysis of estimated breeding values (EBV) of bulls with progeny tests. Since reliabilities are generally in the range of 80–90% for milk production traits, this is equivalent to a GWAS based on individual phenotypes for a trait with a heritability of 80–90%, similar to height in humans. In addition to analysis of individual markers, several studies have analyzed chromosomal segments. This greatly reduces the total number of tests and also reduces the problem that tests on closely linked markers are highly correlated. In this case the different haplotypes present in the population for each segment are determined, and the variance among EBV associated with the different haplotypes is determined (e.g., Cole *et al.*, 2009).

Unlike the situation in most human analyses, the animals analyzed in GWAS are highly related. Thus the genotype effect associated with a specific marker or chromosomal segment will be confounded with the general polygenic variance, because animals with the same allele or haplotype will tend to be related. Thus in addition to the problem of multiple comparisons, it is also necessary to correct for common polygenic effects among relatives. Three basic solutions have been proposed:

1. Inclusion of a relationship matrix based on known relationships or a pseudo relationship matrix based on genotype similarity among the genotyped individuals. Kang *et al.* (2010) developed a program, EMMAX, which includes an identity-by-state matrix based on all genotyped markers in the GWAS model. This program also corrects for multiple comparisons.
2. Determination of marker effects from a model that computes genomic evaluations (Cole *et al.*, 2009). This method also includes incorporation of a genomic relationship matrix in the model but differs from the previous method in that the effects of all valid markers are included in a single analysis.
3. The "a posteriori granddaughter design" (APGD) (Weller and Ron, 2011). This design differs from the original granddaughter design, described in Chapter 7, section "The Granddaughter Design," proposed for analysis by microsatellites, in that with tens of thousands of markers, haplotypes of 50–100 markers can now be determined for each chromosomal segment throughout the entire genome. Due to the huge number of different haplotypes present in outbred populations, nearly all grandsires will be heterozygous for any specific chromosomal region. In this design the problem of confounding due to relationships among animals is solved, because the QTL effects are estimated within grandsire families. This design will be discussed in more detail in section "Determination of Concordance by the 'APGD.'"

Detection of QTN: Is It Worth the Effort?

The arguments for and against QTN determination were summarized by Weller and Ron (2011). The arguments against extending significant effort toward determining QTN can be summarized as follows:

1. The infinitesimal model appears to be approximately accurate for the traits of interest. This corresponds to the results found for human height (Yang *et al.*, 2010) noted previously.
2. Even if a QTN is detected, it may not be useful in selection, either because the economically favorable allele is already at a high frequency in the population or because the net effect of the QTN on the selection index is zero. The favorable allele for *ABCG2* that increases milk protein content and concentration is already at high frequency in all commercial dairy cattle populations (Ron *et al.*, 2006). With respect to *DGAT1* (Grisart *et al.*, 2002; Winter *et al.*, 2002), the allele that increases fat production lowers milk and protein production.
3. Current genomic evaluation methods based on LD are able to achieve reliabilities greater than 0.7 for young animals without records or progeny.
4. Detection of QTN is expensive and time consuming.

The main points in favor of QTN determination are as follows:

1. Once the QTN is determined, this will yield useful information on gene function and QTL architecture. Although the two QTN determined in dairy cattle are both missense mutations, this is not the case for the two other QTN determined in other farm animal species (Ron and Weller, 2007).
2. Understanding the ties between genetic variation and functional characteristics of specific genes may contribute to drug discovery for both farm animals and human.
3. Although SNPs in close linkage to a major QTN will generally display highly significant effects, the effect will still be less than the effect obtained with the QTN (Cohen-Zinder *et al.*, 2005).
4. Population-wide LD relationships change over time, which will reduce the efficiency of genomic selection.
5. Allelic frequencies for a marker in LD will not accurately reflect the allelic frequencies of the QTN.
6. If the QTN is determined, then selection and introgression can also be applied to other populations and breeds, including those populations which do not have effective genomic evaluation programs.

QTN Determination in Farm Animals: What Constitutes Proof?

Glazier *et al.* (2002) noted that the most conclusive evidence that the QTN has been found is a demonstration that replacement of the variant nucleotide results in swapping one phenotypic variant for another. Until recently this was not an option for advanced animal species. With the advent of ZFN, TALEN, and CRISPR/Cas9 technology, this has now become a possibility. Kang *et al.* (2014) reviewed the current status of gene editing technology in farm animals with emphasis on the pig, but ethical objections remain for application to farm animals. Therefore, similar to the situation in humans, all analyses to date have been based on analysis of existing populations.

Considering these limitations, how does one prove that a candidate polymorphism is in fact the QTN? As noted by Mackay (2001), “[t]he only option … is to collect multiple pieces of evidence, no single one of which is convincing, but which together consistently point to a candidate gene.” In the following sections we will consider the main criteria which have been proposed as proof that a QTN has been determined.

Concordance between DNA-Level Genotypes and QTL Status

As noted by Ron and Weller (2007), the most convincing proof of QTN determination in farm animals is "concordance," that is, to demonstrate for a group of animals that their genotypes for the putative QTN correspond to their inferred genotypes for the QTL. Complete concordance is obtained only if:

1. All grandsires homozygous for the QTL are also homozygous for the putative QTN.
2. All grandsires heterozygous for the QTL are also heterozygous for the putative QTN.
3. The same putative QTN allele is associated with the positive QTL allele in all heterozygous sires.

Given these limitations, approximately 20 animals with QTL genotype determined are sufficient to reject the hypothesis of concordance by chance throughout the genome (Ron and Weller, 2007).

Determination of Concordance by the "APGD"

The APGD (Weller and Ron, 2011) not only allows for determination of segregating QTL but unlike other methods proposed for GWAS can also be used to determine QTL genotypes of the grandsires. That is, for a chromosomal region containing a segregating QTL, a significant contrast is expected between the two grandsire haplotypes only if the grandsire is heterozygous for the QTL. Thus the first two criteria for concordance can be tested by the magnitude of the within-family contrast.

Weller *et al.* (2013) applied the APGD to 52 sire families, each with greater than or equal to 100 genotyped sons with genetic evaluations based on progeny tests. The analysis was applied to the autosomal segment with the SNP with the greatest effect in the genomic evaluation of each of 33 traits. The statistical model included the effects of sire and haplotype nested within sire. All traits except for two had a significant within-family haplotype effect.

Weller *et al.* (2014a) applied the APGD to the entire genome. Of 617 haplotype segments spanning the entire bovine genome and each including approximately 5×10^6 bp, 5 cMorgans, and 50 genes, 608 autosomal segments were analyzed. The statistical model of Weller *et al.* (2013) was used for each haplotype segment. For all 33 traits, there was at least one chromosomal region in which the nominal probability for the haplotype effect was less than 10^{-8}, which corresponds to genome-wide significance of less than 10^{-4} (Lander and Kruglyak, 1995). The Manhattan plot for net merit, the main US selection index, is given in Figure 20.1. Net merit had seven chromosomes with nominal probabilities of less than 10^{-8}. For each of those putative QTL, at least one grandsire family had a within-family contrast with a *t*-value of greater than 3.

An updated search for QTL in the Holstein genome was conducted in 2015 using the APGD. The number of Holstein sires with at least 100 genotyped and progeny-tested sons has increased from the previous 52 to 71 for a total of 14,246 sons. The bovine genome was divided into 621 segments of approximately 100 markers each. There were 55 chromosomal regions that met the significance criterion of $P < 10^{-15}$, as compared to 30 regions found by Weller *et al.* (2014a). All traits had at least one significant effect, except for protein yield, daughter stillbirth rate, and four conformation traits.

Confidence intervals (CI) of 90% were determined for all effects by application of a nonparametric bootstrap (Visscher *et al.*, 1996). The length of CI ranged from 2 to 15 chromosomal segments. In all cases, the CI included only part of the chromosome. No significant relationship between log probability of the effect and CI length was found, even though probabilities ranged

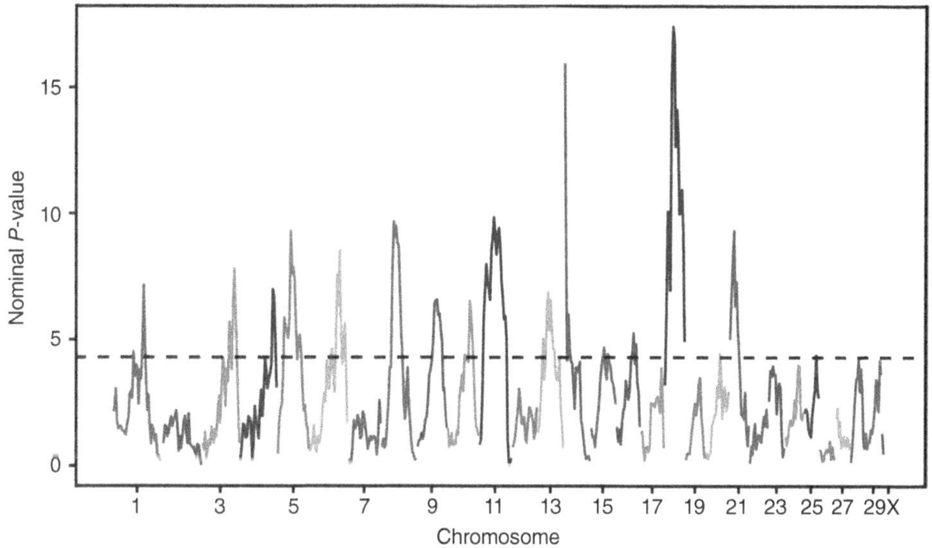

Figure 20.1 Manhattan plot of the a posteriori granddaughter design for net merit. Nominal $P(-\log_{10})$ is plotted as a function of haplotype segment. The dotted line at 4.3 corresponds to genome-wide $P=0.05$.

from 10^{-15} to 10^{-41} on chromosome 3 for protein percentage. At least six of the regions displayed a bimodal effect distribution in the bootstrap analysis, which indicates more than a single QTL segregating on the chromosome.

Results for yield traits were compared with those recently reported for Australian Holsteins, which found effects with nominal probabilities of less than 10^{-20} on six chromosomes (excluding effects on chromosome 14, which clearly result from *DGAT1*) when each SNP effect was estimated as a fixed effect. For US Holsteins, a nominal probability of less than 10^{-6} was found in our study for the same trait in nearly the same chromosomal location, except for the effect of fat percentage on chromosome 27.

Determination of Phase for Grandsires Heterozygous for the QTL

Even with application of the APGD design, it is generally only possible to determine either that the parent is homozygous or heterozygous for the QTL. If the parent is homozygous for the QTL, it will generally not be possible to determine with any certainty if the individual is homozygous for the "positive" or "negative" allele (Israel and Weller, 1998). Thus in testing for concordance generally individuals are scored only as either homozygous or heterozygous.

In order to determine phase of the putative QTN relative to the QTL effects for heterozygous sires, haplotypes for each QTL region can be determined using the PLINK software (http://pngu.mgh.harvard.edu/purcell/plink/; Purcell *et al.*, 2007). For sires for which haplotypes cannot be determined, phase can be determined by genotyping a sample of 5–10 sons for the putative QTN and linked markers. Since only very short chromosomal segments are considered, the frequency of recombination within the segment can be considered zero, which dramatically simplifies haplotype determination.

Determination of Recessive Lethal Genes by GWAS and Effects Associated with Heterozygotes

With GWAS analyses performed on large numbers of individuals, lethal recessives may be discovered from haplotypes that are relatively common in the population, but never appear as homozygous. If the number of genotyped individuals is large, expected numbers of individuals homozygous for a specific haplotype will also be sufficiently large so that the hypothesis that the lack of homozygotes occurred by chance can be rejected with high power. For example, if 1000 individuals are genotyped and the frequency of the haplotype is 0.1, then the expected number of homozygotes based on Hardy–Weinberg equilibrium should be $1000(0.1)^2 = 10$. If 10 homozygotes are expected in the sample, based on the frequency of the haplotype, then the probability of obtaining zero homozygotes by chance based on the assumption of a Poisson distribution is 4.54×10^{-5}. Again, since the entire genome is generally analyzed and each chromosomal segment contains several different haplotypes, significance criteria must account for a large number of multiple comparisons.

Based on this method, VanRaden *et al.* (2011a) discovered five new recessive defects in the US Holstein, Jersey, and Brown Swiss dairy cattle populations. They postulated that the lack of homozygous live births would result in an observed reduction in daughter fertility for heterozygous sires, as measured by daughter pregnancy rate or cow conception rate. In field data an early-term abortion will generally be scored as nonconception. Of 11 haplotypes with sufficient power to reject the hypothesis of zero homozygotes by chance, five also showed a significant reduction of fertility for heterozygous sires. Thus although recessive lethals are generally considered major genes, with respect to their "pleiotropic" effects on fertility, they could be considered QTL.

Determination of the causative polymorphism is simpler in this case, because any polymorphism which is homozygous in live individuals can be rejected. For the three lethals found in Holstein, causative mutations have been identified for HH1 and HH3, but not HH2 (McClure *et al.*, 2014).

Verification of QTN by Statistical and Biological Methods

Concordance alone cannot be considered as proof positive that the QTN has been determined. If a specific chromosomal segment containing the QTN is conserved, then more than a single polymorphism could display complete concordance. However, so far this has not been shown to be the case. Ron and Weller (2007) summarized additional types of evidence, both statistical and biological, that have been used to verify QTN identity. They proposed the following criteria for functional confirmation of a putative QTN:

1. The gene has a known physiological role in the phenotype of the quantitative trait.
2. "Knockout" mutations in this gene affect the trait, even in other species.
3. The gene is preferentially expressed in organs related to the quantitative trait—for example, the mammary gland for milk production traits.
4. The gene is preferentially expressed in developmental stages related to the phenotype—for example, at the onset of lactation for milk production traits.
5. The putative QTN is validated by functional assays, for example, demonstrating either unequal production of the alternative alleles or differences in gene and protein function. For example, expression of recombinant *DGAT1* protein differing only at the K232A mutation demonstrates that this mutation affects the V_{max} of the enzyme in a direction that is in agreement with the observed phenotypic effect (Grisart *et al.*, 2004).

Of the four validated QTN discovered so far in commercial animal populations, none meet all five criteria, but all meet at least two. Statistical methods for QTN validation include demonstrating that:

1. The effect of the putative QTN accounts for the entire effect observed by interval mapping, in this case application of the APGD.
2. No other polymorphisms in LD with the QTL have significant effects in models that also include the effect of the putative QTN.
3. The same QTN is detected in diverse populations.
4. Signatures of selection correspond to the effect associated with the QTN (Glick *et al.,* 2012).

The missense mutation in *ABCG2* apparently meets all four criteria (Cohen-Zinder *et al.,* 2005). The first two criteria may not hold if more than a single polymorphism within the gene affects the gene function. This is apparently the case with *DGAT1*, in which polymorphisms in the promoter also affect fat concentration (Bennewitz *et al.,* 2004a).

Summary

Methods were described to determine and validate QTN for commercial animal species. The "gold standard" used to validate QTN in laboratory organisms, demonstration that replacement of the variant nucleotide results in swapping one phenotypic variant for another, cannot as yet be applied to commercial species. As noted by Mackay (2001), "The only option…is to collect multiple pieces of evidence, no single one of which is convincing, but which together consistently point to a candidate gene." For farm animals the most convincing proof is concordance between QTL genotypes and the causative polymorphism, although additional validation is also necessary.

Although to date only four QTN in farm animals can be considered verified (Ron and Weller, 2007), it is likely that this number will increase dramatically in the near future with complete genome sequencing of large numbers of animals (e.g., Daetwyler *et al.,* 2014).

21 Future Directions and Conclusions

Introduction

Genomic evaluation is still a highly dynamic field, and additional discoveries and new ideas will probably transform the discipline in the next few years. The last 20 years has seen a reduction in genotyping costs from individual markers from approximately $10 per microsatellite genotype in the early 1990s to $0.0005 currently for individual SNP genotypes on high-density BeadChips. At least 11 countries currently have commercial genomic evaluation programs for dairy cattle: Australia, Ireland, New Zealand, France, Germany, the Netherlands, Denmark, Sweden, Finland, the United States, and Canada with programs that began between 2008 and 2011 (Smaragdov, 2013). Two large consortiums have been established: the North American consortium currently with 25,500 bulls that includes the United States, Canada, the United Kingdom, and Italy; and the European consortium with 30,000 bulls that includes UNCEIA (France), Viking Genetics (Denmark, Sweden, and Finland), DHV-VIT (Germany), CRV (the Netherlands, Flanders), Poland, and Spain.

In the next few years complete genome sequencing costs will probably be in the range of a thousand dollars per individual. In the United States and other advanced countries, genotyping of cows for the low-density chip is becoming nearly as routine as milk protein and somatic cell score analysis. On the other hand, progeny testing of large numbers of young bulls is becoming obsolete. AI companies are marketing young bulls for general service, based on their genomic evaluations. In this final chapter we will consider how these changes will affect the future of animal breeding and attempt to look into our crystal ball for future developments, remembering that since the destruction of the temple, prophecy is given only to deaf mutes, the insane, and young children.

More Markers versus More Individuals with Genotypes

By 2015, more than 700,000 Holstein cows and 27,000 US Holstein bulls with genetic evaluations based on daughter records have been genotyped (https://www.cdcb.us/Genotype/cur_density.html). Numbers are significantly smaller for Jerseys and Brown Swiss. Genomic reliabilities of young bulls based only on pedigree and genotypes have reached approximately 70% for milk production traits, as compared to 85–90% after a standard progeny test. Bias is insignificant for Holsteins but may still be a problem for the smaller breeds (Wiggans *et al.*, 2011). Based on simulation studies, it

Genomic Selection in Animals, First Edition. Joel Ira Weller.

is unlikely that increasing numbers of animals genotyped will significantly increase genomic reliabilities. Similarly, as noted in previous chapters (VanRaden *et al.*, 2009), increasing the number of markers beyond the approximately 45,000 usable markers on medium-density chips will have only a minimal effect on accuracy or bias of evaluations.

Computation of Genomic Evaluations for Cow and Female Calves

The most immediate challenge is to incorporate the huge numbers of cow genotypes into routine genomic evaluation systems. Cow genomic evaluations derived by direct incorporation of cow genotypes into two-step methodologies developed for sire analysis result in biased evaluations, due chiefly to the much lower reliabilities of cow evaluations. Wiggans *et al.* (2012b) proposed a double adjustment of genomic cow evaluations to make them comparable to both evaluations of bulls and comparable with traditional evaluations of nongenotyped cows. With these adjustments the regression of genomic cow evaluations on parent averages are nearly equal to unity. Mean reliabilities of approximately 70% for cows genotyped for the Illumina BovineLD BeadChip have already been obtained in the United States (Wiggans *et al.*, 2013).

Börner and Reinsch (2012) found by simulation studies of a population of 100,000 cows that genomic selection is the method of choice when maximizing the genetic gain per year, but genotyping females may not allow for a reduction in overall breeding costs. Thus, the economic justification of large-scale genotyping of females remains questionable.

Improvement of Genomic Evaluation Methods

Although numerous improvements have been proposed in genomic evaluation methods since the first practical methods for genomic evaluation on actual data were tested in 2008, gains in reliabilities have been at best minimal. It is possible that in the next few years single-step methods will become the industry standard, despite the major increase required in computer resources and the very limited gains in accuracy and bias. One major reason is incorporation of cow data, which is straightforward for single-step methods. The Agricultural Research Service of the USDA has attempted to implement single-step methodology for routine analysis of dairy cattle data, but convergence problems were encountered (G. R. Wiggans, personal communication, face-to-face discussion in 2013 in Beltsville, MD).

Bayesian methods that selectively weight specific markers, although applied in several countries, do not seem to be the wave of the future. Recently, Fernando *et al.* (2014) proposed a single-step Bayesian methodology that combines information from both genotyped and ungenotyped animals. The advantages of this method include that it is not necessary to guess the optimal weights for G^{-1} and A_{22}^{-1} as required by the single-step method of Aguilar *et al.* (2010), there is differential shrinkage of marker effects, it is convenient to draw inference on individual marker effects based on the posterior distribution, and empirical results from other experiments as prior information can be readily incorporated into the analysis. However, this model has not as yet been tested on actual data.

So far nearly all methods proposed have considered only additive genetic effects. In the future methods that also consider dominance and epistasis, variance may be applied (e.g., Toro and Varona, 2010; Ober *et al.*, 2011; Wellmann and Bennewitz, 2012; Zeng *et al.*, 2013). To date these models have only been applied to simulated data sets.

Long-Term Considerations

By 2015, genomic evaluation has only been applied at most for 7 years, slightly more than the mean value for a single generation in cattle. Thus genetic changes in commercial populations due to selection so far have not been a problem and have not been seriously addressed to date for genomic selection. Kemper *et al.* (2015) estimated that the age of the QTL mutations for milk production traits in dairy cattle varies from perhaps 2000 to 50,000 generations old. Toro and Varona (2010) found in simulated populations that the efficiency of genomic selection is eroded after a few generations of selection. Over time three major changes will occur that should impact the efficiency of genomic evaluation as currently applied:

1. Frequencies of favorable QTL allele will increase due to selection.
2. New QTLs will "appear" either due to mutation or drift.
3. Population-wide linkage disequilibrium (LD) between markers and QTL will be degraded.

These changes will now be discussed in detail. Favorable alleles of the QTL with the greatest effects will reach fixation. This has apparently occurred for the ABCG2 allele that increases protein concentration in most commercial populations (Cohen-Zinder *et al.*, 2005). The expected change in gene frequency per generation, Δq, resulting from mass selection is computed as follows (Falconer, 1964):

$$\Delta q = iq(1-q)\frac{\alpha}{\sigma_p} \quad (21.1)$$

where i = selection intensity, q = allele frequency prior to selection, σ_p = phenotypic standard deviation, and α = the allele substitution effect. For example, with an overall selection intensity = 1 for a specific trait, $q=0.5$, and $\alpha/\sigma_p=0.2$, then, $\Delta q=0.05$.

Glick *et al.* (2012) analyzed haplotype trends of 917 Israeli Holstein bulls genotyped for the 54K BeadChip born between 1984 and 2008. Of the haplotypes analyzed, 6735 (38%) had nominally significant Δq values ($p<0.05$). Of these haplotypes, 49% increased in frequency. For 7 of the 20 haplotypes with the largest positive ($\Delta q/q\,(1-q)$) values and frequency greater than 0.25, the mean frequency between 2004 and 2008 was greater than 0.9. These loci will not be useful for further selection, as the economically favorable allele has nearly reached fixation.

Although mutation rates are so low that their effect can be considered negligible over the next century, "new" QTL will appear due to the increase of the frequencies of rare positive alleles in the population. That is, the frequency of very rare positive allele will hardly be affected by current selection strategies, as can be seen from Equation (21.1). However, if the frequencies of these alleles do increase, either due to selection or random drift, then Δq values will increase. This is one explanation why heritabilities have not decreased for milk production traits despite 50 years of selection (deKoning and Weller, 1994). Assuming that the effect of a QTL is additive and only two effective alleles are segregating in the population, then the variance due to the QTL will be $2(1-q)q\alpha^2$. Similar to Equation (21.1) this expression is maximum with $q=0.5$. As the frequency of a rare allele increases, genetic variance due to the locus also increases.

Population-wide LD relationships between markers and segregating QTL will change. Since selection is on the marker alleles, and not directly on the QTL, and since LD tends to decay over generations, the effectiveness of genomic selection based on LD relationships averaged over bulls born over an extended time period should decline.

Bastiaansen *et al.* (2012) evaluated three methods to compute GEBV: the standard multistep method (VanRaden, 2008), a Bayesian method, and a partial least squares regression method. The reference population consisted of 500 genotyped individuals, with all individuals from a single generation, or 100 individuals from each of four generations. Differences in long-term selection response were small. Under selection, applying the first method led to lower inbreeding and a smaller reduction of genetic variance, while a similar response to selection was achieved. After 10 generations of selection, all methods of genomic evaluation gave accuracies in the range of 5–15%. Thus, for genomic selection to be practical continuous reevaluation of marker effects over time is required.

Kemper *et al.* (2012) investigated by simulation if "genotype building" is an appropriate strategy for long-term selection response for a complex trait in dairy cattle. Plant breeders often use a targeted strategy to build a predefined genotype, based on techniques such as backcrossing and gene stacking or pyramiding to introgress desirable genes into an elite variety. They found that an ideal genotype was difficult to achieve, even under simulation conditions, and concluded that selection on overall GEBV with a constraint on coancestry is the most flexible selection strategy. Similarly, 30 years ago Weller and Soller (1981) found that the optimal strategy for combining several loci into a single strain was random mating and mass selection on those individuals with the greatest number of desirable alleles.

Weighting Evaluations of Old versus Young Bulls

As all young sires produced by AI studs now have genomic evaluations, the young bulls with the highest evaluations are now released for general service, while there is little incentive to progeny test young bulls with inferior evaluations. Thus the number of AI sires produced per year with genetic evaluations based on progeny records will decrease. With current multistep genomic evaluation methods, the evaluations of bulls are weighted only with respect to the accuracy of the evaluation, that is, the effective number of daughters. Thus older bulls are generally given greater weights. From the considerations given in the previous section, young bulls, which reflect more accurately the current genomic status of the population, should be given greater weight. The fact that older bulls do not accurately reflect the current genomic status of the population will become more problematic in the future, as the fraction of relatively young bulls out of total bulls with genotypes decreases. So far this problem has not been addressed by current genomic evaluation methods, which assume that QTL effects associated with SNPs are constant over time. The fact that older bulls do not accurately reflect the current genomic status of the population might explain the results of Lourenco *et al.* (2014a) who found that decreasing the number of generations included in the pedigree may result in more accurate genomic evaluations. This problem will also not be solved by single-step methodologies that also give equal weight to all SNPs regardless of the animals' ages.

Direct Genetic Manipulation in Farm Animals

In the previous chapter we considered methods to determine the actual polymorphisms responsible for the observed genetic variation in quantitative traits. We noted that once the specific quantitative trait variants have been identified, these factors can then be included in breeding programs as fixed effects. With the advent of ZFN, TALEN, and CRISPR/Cas9 technology, the possibility also exists to actually change genotypes of embryos. This technology has already been extensively applied to model organisms, such as *Drosophila* (Bassett *et al.*, 2013). Kang *et al.* (2014) reviewed the current status of gene editing technology in farm animals with emphasis on the pig. Using ZFN, TALEN, and CRISPR/Cas9, the epigenome of a locus of interest can now be modified to alter regulatory pathways, which lead eventually to dramatic changes of the phenotype of the animal. However, at present there is significant ethical opposition to application of these technologies to farm animals.

Velogenetics: The Synergistic Use of MAS and Germ-Line Manipulation

In this final section, we will consider combination of genomic evaluation with germ-line manipulation, as first proposed by Georges and Massey (1991) for MAS. Spontaneous oocyte maturation and ovulation do not begin until puberty. For cattle this is at the age of close to 1 year. However, waves of oocyte growth are seen even *in utero*. Activation of primordial follicles starts at 140 days of gestation. Georges and Massey (1991) considered the theoretical possibility to grow, mature, and fertilize prepubertal oocytes *in vitro*. This procedure could possibly reduce the generation interval of cattle to as little as 3–6 months, as compared to the normal biological minimum of close to 2 years. By using *in vitro* fertilization of fetal oocytes by selected progeny-tested sires, annual responses in milk yield could be doubled compared to conventional progeny testing. They term this procedure "velogenetics" and propose the following breeding scheme, updated for genomic evaluation:

1. Selection of "bull granddams" based on genomic evaluations.
2. Selection of fetal "bull dams" based on genomic evaluations.
3. *In vitro* fertilization of fetal oocytes with semen of elite sires.
4. Selection among juvenile male calves based on genomic evaluations.
5. Selected young sires at the age of 1–2 years are mated to cows of commercial population.

Step 3 of this protocol is not possible at present, but until very recently, it was also considered impossible to clone mature mammals and to find QTN for mammalian species.

Summary

In this final chapter, we attempted to look into the future and elaborated on the challenges that genomic selection will face over the next generation and briefly considered how methodologies for genomic evaluation can be adapted to meet these challenges. In the final sections of this chapter we considered expected technological advances which may have major effects on future breeding programs, including new techniques for direct genetic manipulation in farm animals, and *in vitro* fertilization of fetal oocytes. As usual, the story has just begun.

References

Aguilar, I., Misztal, I., Johnson, D. L., Legarra, A., Tsuruta, S., and Lawlor, T. J. (2010) Hot topic: A unified approach to utilize phenotypic, full pedigree, and genomic information for genetic evaluation of Holstein final score. *J. Dairy Sci.* **93**: 743–752.

Anderson, E. C. and Garza, J. C. (2006) The power of single-nucleotide polymorphisms for large-scale parentage inference. *Genetics* **172**: 2567–2582.

Andersson-Elkund, L., Danell, B., and Rendel, J. (1990) Associations between blood groups, blood protein polymorphisms and breeding values for production traits in Swedish Red & White dairy bulls. *Anim. Genet.* **21**: 361–376.

Ashwell, M. S. and Van Tassell, C. P. (1999) Detection of putative loci affecting milk, health, and type traits in a US Holstein population using 70 microsatellite markers in a genome scan. *J. Dairy Sci.* **82**: 2497–2502.

Ashwell, M. S., Rexroad, C. E., Miller, R. H., and VanRaden, P. M. (1996) Mapping economic trait loci for somatic cell score in Holstein cattle using microsatellite markers and selective genotyping. *Anim. Genet.* **27**: 235–242.

Ashwell, M. S., Rexroad, C. E., Miller, R. H., VanRaden, P. M., and Da, Y. (1997) Detection of loci affecting milk production and health traits in an elite US Holstein population using microsatellite markers. *Anim. Genet.* **28**: 216–222.

Ashwell, M. S., Da, Y., VanRaden, P. M., Rexroad, C. E., and Miller, R. H. (1998a) Detection of putative loci affecting conformational type traits in an elite population of United States Holsteins using microsatellite markers. *J. Dairy Sci.* **81**: 1120–1125.

Ashwell, M. S., Da, Y., Van Tassell, C. P., VanRaden, P. M., Miller, R. H., and Rexroad, C. E. (1998b) Detection of putative loci affecting milk production and composition, health, and type traits in a United States Holstein population. *J. Dairy Sci.* **81**: 3309–3314.

Ashwell, M. S., Van Tassell, C. P., and Sonstegard, T. S. (2001) A genome scan to identify quantitative trait loci affecting economically important traits in a US Holstein population. *J. Dairy Sci.* **84**: 2535–2542.

Ashwell, M. S., Heyen, D. W., Sonstegard, T. S., Van Tassell, C. P., Da, Y., VanRaden, P. M., Ron, M., Weller, J. I., and Lewin, H. A. (2004) Detection of quantitative trait loci affecting milk production, health, and reproductive traits in Holstein cattle. *J. Dairy Sci.* **87**: 468–475.

Baruch, E. and Weller, J. I. (2008) Incorporation of discrete genotype effects for multiple genes into animal model evaluations when only a small fraction of the population has been genotyped. *J. Dairy Sci.* **91**: 4365–4371.

Baruch, E. and Weller, J. I. (2009) Incorporation of genotype effects into animal model evaluations when only a small fraction of the population has been genotyped. *Animal* **3**: 16–23.

Baruch, E., Weller, J. I., Cohen-Zinder, M., Ron, M., and Seroussi, E. (2006) Efficient inference of haplotypes from genotypes on a large animal pedigree. *Genetics* **172**: 1757–1765.

Bassett, A. R., Tibbit, C., Ponting, C. P., and Liu, J.-L. (2013) Highly efficient targeted mutagenesis of *Drosophila* with the CRISPR/Cas9 system. *Cell Reports* **4**: 220–228.

Bastiaansen, J. W. M., Coster, A., Calus, M. P. L., van Arendonk, J. A. M., and Bovenhuis, H. (2012) Long term response to genomic selection: Effects of estimation method and reference population structure for different genetic architectures. *Genet. Sel. Evol.* **44**: 3.

Beavis, W. D. (1994) The power and deceit of QTL experiments: Lessons for comparative QTL studies. *Ann. Corn Sorghum Res. Conf. Washington, DC* **49**: 252–268.

Beckmann, J. S. and Soller, M. (1983) Restriction fragment length polymorphisms in genetic improvement: Methodologies, mapping and costs. *Theor. Appl. Genet.* **67**: 35–43.

Benjamini, Y. and Hochberg, Y. (1995) Controlling the false discovery rate: A practical and powerful approach to multiple testing. *J. R. Stat. Soc. Ser. B.* **57**: 289–300.

Bennewitz, J., Reinsch, N., Grohs, C., Leveziel, H., Malafosse, A., Thomsen, H., Xu, N. Y., Looft, C., Kuhn, C., Brockmann, G. A., Schwerin, M., Weimann, C., Hiendleder, S., Erhardt, G., Medjugorac, I., Russ, I., Forster, M., Brenig, B., Reinhardt, F., Reents, R., Averdunk, G., Blumel, J., Boichard, D., and Kalm, E. (2003) Combined analysis of data from two granddaughter designs: A simple strategy for QTL confirmation and increasing experimental power in dairy cattle. *Genet. Sel. Evol.* **35**: 319–338.

Bennewitz, J., Reinsch, N., Paul, S., Looft, C., Kaupe, B.,Weimann, C., Erhardt, G., Thaller, G., Kuhn, C., Schwerin, M., Thomsen, H., Reinhardt, F., Reents, R., and Kalm, E. (2004a) The DGAT1K232A mutation is not solely responsible for the milk production quantitative trait locus on the bovine chromosome 14. *J. Dairy Sci.* **87**: 431–442.

Bennewitz, J., Reinsch, N., Reinhardt, F., Liu, Z., and Kalm, E. (2004b) Top down preselection using marker assisted estimates of breeding values in dairy cattle. *J. Anim. Breed. Genet.* **121**: 307–318.

Blott, S., Kim, J. J., Moisio, S., Schmidt-Kuntzel, A., Cornet, A., Berzi, P., Cambisano, N., Ford, C., Grisart, B., Johnson, D., Karim, L., Simon, P., Snell, R., Spelman, R., Wong, J., Vilkki, J., Georges, M., Farnir, F., and Coppieters, W. (2003) Molecular dissection of a quantitative trait locus: A phenylalanine-to-tyrosine substitution in the transmembrane domain of the bovine growth hormone receptor is associated with a major effect on milk yield and composition. *Genetics* **163**: 253–266.

Boichard, D., Grohs, C., Bourgeois, F., Cerqueira, F., Faugeras, R., Neau, A., Rupp, R., Amigues, Y., Boscher, M. Y., and Leveziel, H. (2003) Detection of genes influencing economic traits in three French dairy cattle breeds. *Genet. Sel. Evol.* **35**: 77–101.

Boichard, D., Fritz, S., Rossignol, M. N., Guillaume, F., Colleau, J. J., and Druet, T. (2006) Implementation of marker-assisted selection: Practical lessons from dairy cattle. *Proceedings of the 8th World Congress Genetics Applied Livestock Production*, Belo Horizonte, MG, Brazil, **22**: 11.

Börner, V. and Reinsch, N. (2012) Optimising multistage dairy cattle breeding schemes including genomic selection using decorrelated or optimum selection indices. *Genet. Sel. Evol.* **44**: 1.

Botstein, D., White, R. L., Skolnick, M., and Davis, R. W. (1980) Construction of a genetic linkage map in man using restriction fragment length polymorphisms. *Am. J. Hum. Genet.* **32**: 314–331.

Bouquet, A. and Juga, J. (2013) Integrating genomic selection into dairy cattle breeding programmes: A review. *Animal* **7**: 705–713.

Bovenhuis, H. and Van Arendonk, J. A. M. (1991) Estimation of milk protein gene frequencies in crossbred cattle by maximum likelihood. *J. Dairy Sci.* **74**: 2728–2736.

Bovenhuis, H. and Weller, J. I. (1994) Mapping and analysis of dairy cattle quantitative trait loci by maximum likelihood methodology using milk protein genes as genetic markers. *Genetics* **137**: 267–280.

Boveri, T. H. (1902) Über mehrpolige Mitosen als Mittel zur Analyse des Zellkerns [Via multi-pole mitoses as a means of analysis of the cell nucleus]. *Verh.d. Phys.-Med. Ges.Würzburg N.F.* **35**: 67–90.

Bovine Genome Sequencing and Analysis Consortium, Elsik, C. G., *et al.* (2009) The genome sequence of taurine cattle: A window to ruminant biology and evolution. *Science* **324**: 522–528.

Brascamp, E. W. (1984) Selection indices with constraints. *Anim. Breed. Abs.* **52**: 645–654.

Bredbacka, P. and Koskinen, M. T. (1999) Microsatellite panels suggested for parentage testing in cattle: Informativeness revealed in Finnish Ayrshire and Holstein Friesian populations. *Agric. Food Sci. Finland* **8**: 233–237.

Brookes, A. J. (1999) The essence of SNPs. *Gene* **234**: 177–186.

Browning, S. R. and Browning, B. L. (2007) Rapid and accurate haplotype phasing and missing-data inference for whole-genome association studies by use of localized haplotype clustering. *Am. J. Hum. Genet.* **81**: 1084–1097.

Buch, L. H., Sørensen, M. K., Berg, P., Pedersen, L. D., and Sørensen, A. C. (2012) Genomic selection strategies in dairy cattle: Strong positive interaction between use of genotypic information and intensive use of young bulls on genetic gain. *J. Anim. Breed. Genet.* **129**: 138–151.

de los Campos, G., Naya, H., Gianola, D., Crossa, J., Legarra, A., Manfredi, E., Weigel, K., and Cotes, J. M. (2009) Predicting quantitative traits with regression models for dense molecular markers and pedigree. *Genetics* **182**: 375–385.

Chen, J., Liu, Z., Reinhardt, F., and Reents, R. (2011) Reliability of genomic prediction using imputed genotypes for German Holsteins: Illumina 3K to 54K bovine chip. *Interbull Bull.* **44**: 51–54.

Cheverud, J. M. (2001) A simple correction for multiple comparisons in interval mapping genome scans. *Heredity* **87**: 52–58.

Christensen, L. G., Madsen, P., and Petersen, J. (1982) The influence of incorrect sire-identification on the estimates of genetic parameters and breeding values. *Proceedings of the 2nd World Congress on Genetics Applied to Livestock Production*, Madrid, Spain, **7**: 200–208.

Churchill, G. A. and Doerge, R. W. (1994) Empirical threshold values for quantitative trait mapping. *Genetics* **138**: 963–971.

Cleveland, M. A., Hickey, J. M., and Forni, S. (2012) A common dataset for genomic analysis of livestock populations. *G3* **2**: 429–435.

Cohen, M., Seroussi, E., Ron, M., Reichenstein, M., Plis-Finarov, A., Shani, M., and Weller, J. I. (2002) Population-wide linkage disequilibrium between a SNP and a QTL affecting milk protein production on BTA6 in dairy cattle. *Proceedings of the 7th World Congress Genetics Applied to Livestock Production*, Montpellier, France, **31**: 51–54.

Cohen-Zinder, M., Seroussi, E., Larkin, D. M., Loor, J. J., Everts-van der Wind, A., Lee, J. H., Drackley, J. K., Band, M. R., Hernandez, A. G., Shani, M., Lewin, H. A., Weller, J. I., and Ron, M. (2005) Identification of a missense mutation in the bovine *ABCG2* gene with a major effect on the QTL on chromosome 6 affecting milk yield and composition in Holstein cattle. *Genome Res.* **15**: 936–944.

Cole, J. B., VanRaden, P. M., O'Connell, J. R., Van Tassell, C. P.Sonstegard, T. S., Schnabel, R. D., Taylor, J. F., and Wiggans, G. R. (2009) Distribution and location of genetic effects for dairy traits. *J. Dairy Sci.* **92**: 2931–2946.

Colombani, C., Legarra, A., Fritz, S., Guillaume, F., Croiseau, P., Ducrocq, V., and Robert-Granie, C. (2013) Application of Bayesian least absolute shrinkage and selection operator (LASSO) and Bayes Cπ methods for genomic selection in French Holstein and Montbeliarde breeds. *J. Dairy Sci.* **96**: 575–591.

Cowan, C. M., Dentine, M. R., and Colye, T. (1992) Chromosome substitution effects associated with k-Casein and b-Lactoglobulin in Holstein cattle. *J. Dairy Sci.* **75**: 1097–1104.

Cunningham, E. P. (1969) The relative efficiencies of selection indexes. *Acta Agri. Scand.* **19**: 45–48.

Daetwyler, H. D., Villanueva, B., Bijma, P., and Woolliams, J. A. (2007) Inbreeding in genome-wide selection. *J. Anim. Breed. Genet.* **124**: 369–376.

Daetwyler, H. D., Villanueva, B., and Woolliams, J. A. (2008) Accuracy of predicting the genetic risk of disease using a genome-wide approach. *PLoS One* **3**: e3395.

Daetwyler, H. D., Capitan, A., Pausch, H., *et al.* (2014) Whole-genome sequencing of 234 bulls facilitates mapping of monogenic and complex traits in cattle. *Nat. Genet.* **46**: 858–865.

Darvasi, A., Vinreb, A., Minke, V., Weller, J. I., and Soller, M. (1993) Detecting marker-QTL linkage and estimating QTL gene effect and map location using a saturated genetic map. *Genetics* **134**: 943–951.

Dekkers, J. C. M. (2007) Prediction of response to marker-assisted and genomic selection using selection index theory. *J. Anim. Breed. Genet.* **124**: 331–341.

deKoning, G. J. and Weller, J. I. (1994) Efficiency of direct selection on quantitative trait loci for a two-trait breeding objective. *Theor. Appl. Genet.* **88**: 669–677.

deRoos, A. P. W., Schrooten, C., Veerkamp, R. F., and Van Arendonk, J. A. M. (2011) Effects of genomic selection on genetic improvement, inbreeding, and merit of young versus proven bulls. *J. Dairy Sci.* **94**: 1559–1567.

Druet, T. and Georges, M. (2010) A hidden Markov Model combining linkage and linkage disequilibrium information for haplotype reconstruction and quantitative trait locus fine mapping. *Genetics* **184**: 789–798.

Edwards, M. D., Stuber, C. W., and Wendel, J. F. (1987) Molecular-marker-facilitated investigations of quantitative-trait loci in maize. I. Numbers, genomic distribution and types of gene action. *Genetics* **116**: 113–125.

Efron, B. and Tibshirani, R. J. (1993) *An Introduction to the Bootstrap*. Chapman & Hall, New York, NY.

El-Kassaby, Y. A. and Lstibůrek, M. (2009) Breeding without breeding. *Genet. Res. (Camb.)* **91**: 111–120.

Eshed, Y. and Zamir, D. (1996) Less-than-additive epistatic interactions of quantitative trait loci in tomato. *Genetics* **143**: 1807–1917.

Falconer, D. S. (1964) *Introduction to Quantitative Genetics*. Oliver and Boyd, Edinburgh.

Farnir, F., Coppieters, W., Arranz, J. J., Berzi, P., Cambisano, N., Grisart, B., Karim, L., Marcq, F., Moreau, L., Mni, M., Nezer, C., Simon, P., Vanmanshoven, P., Wagenaar, D., and Georges, M. (2000) Extensive genome-wide linkage disequilibrium in cattle. *Genome Res.* **10**: 220–227.

Fernando, R. L. and Grossman, M. (1989) Marker assisted selection using best linear unbiased prediction. *Genet. Sel. Evol.* **21**: 467–477.

Fernando, R. L., Nettleton, D., Southey, B. R., Dekkers, J. C. M., Rothschild, M. F., and Soller, M. (2004) Controlling the proportion of false positives in multiple dependent tests. *Genetics* **166**: 611–619.

Fernando, R. L., Dekkers, J. C. M., and Garrick, D. J. (2014) A class of Bayesian methods to combine large numbers of genotyped and non-genotyped animals for whole-genome analyses. *Genet. Sel. Evol.* **46**: 50.

Fisher, R. A. (1918) The correlation between relatives on the supposition of Mendelian inheritance. *Trans. Royal Soc. Edinburgh* **52**: 399–433.

Freeman, J. L., Perry, G. H., Feuk, L.Redon, R., McCarroll, S. A., Altshuler, D. M., Aburatani, H., Jones, K. W., Tyler-Smith, C., Hurles, M. E., Carter, N. P., Scherer, S. W., and Lee, C. (2006) Copy number variation: New insights in genome diversity. *Genome Res.* **16**: 949–961.

Garrick, D. J. (2007) Equivalent mixed model equations for genomic selection. *J. Dairy Sci.* **90**(Suppl. 1): 376.

Geldermann, H., Pieper, U., and Weber, W. E. (1986) Effect of misidentification on the estimation of breeding value and heritability in cattle. *J. Anim. Sci.* **63**: 1759–1768.

Georges, M. and Massey, J. M. (1991) Velogenetics, or the synergistic use of marker assisted selection and germ-line manipulation. *Theriogenology* **35**: 151–159.

Georges, M., Nielsen, D., Mackinnon, M., Mishra, A., Okimoto, R., Pasquino, A. T., Sargent, L. S., Sorensen, A., Steele, M. R., Zhao, X., Womack, J. E., and Hoeschele, I. (1995) Mapping quantitative trait loci controlling milk production in dairy cattle by exploiting progeny testing. *Genetics* **139**: 907–920.

Gianola, D., de los Campos, G. A., Hill, W. G., Manfredi, E., and Fernando, R. L. (2009) Additive genetic variability and the Bayesian alphabet. *Genetics* **183**: 347–363.

Gibson, J. P. (1994) Short-term gain at the expense of long-term response with selection on identified loci. *Proceedings of the 5th World Congress Genetics Applied Livestock Production*, Guelph, ON, **21**: 201–204.

Glazier, A. M., Nadeau, J. H., and Aitman, T. J. (2002) Finding genes that underlie complex traits. *Science* **298**: 2345–2349.

Glick, G.Shirak, A., Uliel, S., Zeron, Y., Ezra, E., Seroussi, E., Ron, M., and Weller, J. I. (2012) Signatures of contemporary selection in the Israeli Holstein dairy cattle. *Anim. Genet.* **43**(Suppl. 1): 45–55.

Glowatzki-Mullis, M. L., Gaillard, C., Wigger, G., and Fries, R. (1995) Microsatellite-based parentage control in cattle. *Anim. Genet.* **26**: 7–12.

Goddard, M. E. (1992) A mixed model for analyses of data on multiple genetic markers. *Theor. Appl. Genet.* **83**: 878–886.

Goddard, M. E. and Hayes, B. J. (2007) Genomic selection. *J. Anim. Breed. Genet.* **124**: 323–330.

Goddard, M. E., Hayes, B., McPartlan, H., Chamberlain, A. J. (2006) Can the same genetic markers be used in multiple breeds? *Proceedings of the 8th World Congress Genetics Applied Livestock Production*, Belo Horizonte, Brazil. 22–14.

Gondro, C., van der Werf, J., and Hayes, B. (2013) *Genome-Wide Association Studies and Genomic Prediction*. Humana Press, New York, NY.

Gorbach, D. M., Makgahlela, M. L., Reecy, J. M., Kemp, S. J., Baltenweck, I., Ouma, R., Mwai, O., Marshall, K., Murdoch, B., Moore, S., and Rothschild, M. F. (2010) Use of SNP genotyping to determine pedigree and breed composition of dairy cattle in Kenya. *J. Anim. Breed. Genet.* **127**: 348–351.

Gotz, K. U. and Ollivier, L. (1992) Theoretical aspects of applying sib-pair linkage tests to livestock species. *Genet. Sel. Evol.* **24**: 29–42.

Grisart, B., Coppieters, W., Farnir, F., Karim, L., Ford, C., Berzi, P., Cambisano, N., Mni, M., Reid, S., Simon, P., Spelman, R., Georges, M., and Snell, R. (2002) Positional candidate cloning of a QTL in dairy cattle: Identification of a missense mutation in the bovine DGAT1gene with major effect on milk yield and composition. *Genome Res.* **12**: 222–231.

Grisart, B., Farnir, F., Karim, L., Cambisano, N., Kim, J. J., Kvasz, A., Mni, M., Simon, P., Frere, J. M., Coppieters, W., and Georges, M. (2004) Genetic and functional confirmation of the causality of the DGAT1K232A quantitative trait nucleotide in affecting milk yield and composition. *Proc. Natl. Acad. Sci. U. S. A.* **101**: 2398–2403.

Grodzicker, T., Williams, J., Sharp, P., and Sambrook, J. (1974) Physical mapping of temperature sensitive mutations of adenoviruses. *Cold Spring Harb. Symp. Quant. Biol.* **39**: 439–446.

Guillaume, F., Fritz, F., Boichard, D., and Druet, T. (2008) Estimation by simulation of the efficiency of the French marker-assisted selection program in dairy cattle. *Genet. Sel. Evol.* **40**: 91–102.

Habier, D., Fernando, R. L., and Dekkers, J. C. M. (2009) Genomic selection using low-density marker panels. *Genetics* **182**: 343–353.

Habier, D., Fernando, R. L., Kizilkaya, K., and Garrick, D. J. (2011) Extension of the Bayesian alphabet for genomic selection. *BMC Bioinf.* **12**: 186.

Haldane, J. B. S. (1919) The combination of linkage values and the calculation of distances between the loci of linked factors. *J. Genet.* **8**: 299.

Haley, C. S. and Knott, S. A. (1992) A simple regression method for mapping quantitative trait loci in line crosses using flanking markers. *Heredity* **69**: 315–324.

Hardy, G. H. (1908) Mendelian proportions in mixed population. *Science* **28**: 49.

Haseman, J. K. and Elston, R. C. (1972) The investigation of linkage between a quantitative trait and a marker locus. *Behav. Genet.* **2**: 3–19.

Hayes, B. and Goddard, M. E. (2001) The distribution of the effects of genes affecting quantitative traits in livestock. *Genet. Sel. Evol.* **33**: 209–229.

Hazel, L. N. (1943) The genetic basis for constructing selection indexes. *Genetics* **28**: 476–490.

Heaton, M. P., Harhay, G. P., Bennett, G. L., Stone, R. T., Grosse, W. M., Casas, E., Keele, J. W., Smith, T. P., Chitko-McKown, C. G., and Laegreid, W. W. (2002) Selection and use of SNP markers for animal identification and paternity analysis in U.S. beef cattle. *Mamm. Genome* **13**: 272–281.

Henderson, C. R. (1973) Sire evaluation and genetic trends. In *Proceedings of the Animal Breeding and Genetics Symposium in Honor of Dr. D. L. Lush*. ASAS and ADSA, Champaign, IL. pp. 10–41.

Henderson, C. R. (1976) A simple method for the inverse of a numerator relationship matrix used in prediction of breeding values. *Biometrics* **32**: 69–83.

Henderson, C. R. (1984) *Applications of Linear Models in Animal Breeding*. University of Guelph, Guelph, ON.

Heyen, D. W., Weller, J. I., Ron, M., Band, M., Beever, J. E., Feldmesser, E., Da, Y., Wiggans, G. R., VanRaden, P. M., and Lewin, H. A. (1999) A genome scan for QTL influencing milk production and health traits in dairy cattle. *Physiol. Genomics* **1**: 165–175.

Hill, W. G. (1971) Investment appraisal for national breeding programmes. *Anim. Prod.* **13**: 37–50.

Hill, W. G., Salisbury, B. A., and Webb, A. J. (2008) Parentage identification using single nucleotide polymorphism genotypes: Application to product tracing. *J. Anim. Sci.* **86**: 2508–2517.

Hoeschele, I. and VanRaden, P. M. (1993a) Bayesian analysis of linkage between genetic markers and quantitative trait loci. I. Prior knowledge. *Theor. Appl. Genet.* **85**: 953–960.

Hoeschele, I. and VanRaden, P. M. (1993b) Bayesian analysis of linkage between genetic markers and quantitative trait loci. II. Combining prior knowledge with experimental evidence. *Theor. Appl. Genet.* **85**: 946–952.

Holmberg, M. and Andersson-Eklund, L. (2004) Quantitative trait loci affecting health traits in Swedish dairy cattle. *J. Dairy Sci.* **87**: 2653–2659.

Hudson, R. R. (1985) The sampling distribution of linkage disequilibrium under an infinite alleles model without selection. *Genetics* **109**: 611–631.

Ihara, N., Takasuga, A., Mizoshita, K., Takeda, H., Sugimoto, M., Mizoguchi, Y., Hirano, T., Itoh, T., Watanabe, T., Reed, K. M., Snelling, W. M., Kappes, S. M., Beattie, C. W., Bennett, G. L., and Sugimoto, Y. (2004) A comprehensive genetic map of the cattle genome based on 3802 microsatellites. *Genome Res.* **14**: 1987–1998.

Israel, C. and Weller, J. I. (1998) Estimation of candidate gene effects in dairy cattle populations. *J. Dairy Sci.* **81**: 1653–1662.

Israel, C. and Weller, J. I. (2000) Effect of misidentification on genetic gain and estimation of breeding value in dairy cattle populations. *J. Dairy Sci.* **83**: 181–187.

James, J. W. (1982) Construction, uses and problems of multitrait selection indices. *Proceedings of the Second World Congress on Genetics Applied to Livestock Production*, Madrid, Spain. **5**: 130–139.

Jiménez-Montero, J. A., Gonzalez-Recio, O., and Alenda, R. (2012) Comparison of methods for the implementation of genome-assisted evaluation of Spanish dairy cattle. *J. Dairy Sci.* **96**: 625–634.

Johanssen, W. (1903) *Uber Erblichkeit in Polulationen und in reinen Linien [About heritability in populations and in pure lines]*. G. Fisher, Jena.

Kahler, A. L. and Wherhahn, C. F. (1986) Association between quantitative traits and enzyme loci in the F2 population of a maize hybrid. *Theor. Appl. Genet.* **72**: 15–26.

Kang, H. M., Sul, J. J., Service, S. K., Zaitlen, N. A., Kong, S.-Y., Freimer, N. B., Sabatti, C., and Eskin, E. (2010) Variance component model to account for sample structure in genome-wide association studies. *Nat. Genet.* **42**: 348–354.

Kang, Q., Hu, Y., Zou, Y., Hu, W., Li, L., Chang, F., Li, Y., Lu, D., Sun, Z., Zhang, R., Hu, X., Li, Q., Dai, Y., and Li, N. (2014) Improving pig genetic resistance and muscle production through molecular biology. *Proceedings of the 10th World Congress on Genetics Applied to Livestock Production*, Vancouver, BC. Accessed December 9, 2015. https://www.researchgate.net/publication/267156847_Improving_pig_genetic_resistance_and_muscle_production_through_molecular_biology

Kemper, K. E., Bowman, P. J., Pryce, J. E., Hayes, B. J., and Goddard, M. E. (2012) Long-term selection strategies for complex traits using high-density genetic markers. *J. Dairy Sci.* **95**: 4646–4656.

Kemper, K. E., Hayes, B. J., Daetwyler, H. D., and Goddard, M. E. (2015) How old are quantitative trait loci and how widely do they segregate? *J. Anim. Breed. Genet.* **132**: 121–134.

Khatkar, M. S., Thomson, P. C., Tammen, I., and Raadsma, H. W. (2004) Quantitative trait loci mapping in dairy cattle: Review and meta-analysis. *Genet. Sel. Evol.* **36**: 163–190.

Kizilkaya, K., Fernando, R. L., and Garrick, D. J. (2010) Genomic prediction of simulated multibreed and purebred performance using observed fifty thousand single nucleotide polymorphism genotypes. *J. Anim. Sci.* **88**: 544–551.

Klungland, H., Sabry, A., Heringstad, B., Olsen, H. G., Gomez-Raya, L., Vage, D. I., Olsaker, I., Odegard, J., Klemetsdal, G., Schulman, N., Vilkki, J., Ruane, J., Aasland, M., Ronningen, K., and Lien, S. (2001) Quantitative trait loci affecting clinical mastitis and somatic cell count in dairy cattle. *Mamm. Genome* **12**: 837–842.

Knott, S. A., Elsen, J. M., and Haley, C. S. (1994) Multiple marker mapping of quantitative trait loci in half-sib populations. *Proceedings of the 4th World Congress Genetics Applied Livestock Production*, Guelph, ON, **21**, 33–36.

Knott, S. A., Elsen, J. M., and Haley, C. S. (1996) Methods for multiple-marker mapping of quantitative trait loci in half-sib populations. *Theor. Appl. Genet.* **93**: 71–80.

Kolbehdari, D., Robinson, J. A. B., and Schaeffer, L. R. (2006) Mapping of QTL using regression on multiple marker haplotypes. *Proceedings of the 8th World Congress on Genetics Applied to Livestock Production*, Belo Horizonte, MG. 20–15.

Kruglyak, L. and Lander, E. S. (1995) High-resolution genetic mapping of complex traits. *Am. J. Hum. Genet.* **56**: 1212–1223.

Kuhn, C., Bennewitz, J., Reinsch, N., Thomsen, H., Looft, C., Brockmann, G. A., Schwerin, M., Weimann, C., Hiendieder, S., Erhardt, G., Medjugorac, I., Forster, M., Brenig, B., Reinhardt, F., Reents, R., Russ, I., Averdunk, G., Blumel, J., and Kalm, E. (2003) Quantitative trait loci mapping of functional traits in the German Holstein cattle population. *J. Dairy Sci.* **86**: 360–368.

Kuhn, C., Thaller, G., Winter, A., Bininda-Emonds, O. R. P., Kaupe, B., Erhardt, G., Bennewitz, J., Schwerin, M., and Fries, R. (2004) Evidence for multiple alleles at the DGAT1 locus better explains a quantitative trait locus with major effect on milk fat content in cattle. *Genetics* **167**: 1873–1881.

Lande, R. and Thompson, R. (1990) Efficiency of marker-assisted selection in the improvement of quantitative traits. *Genetics* **124**: 743–756.

Lander, E. S. and Botstein, D. (1989) Mapping Mendelian factors underlying quantitative traits using RFLP linkage maps. *Genetics* **121**: 185–199.

Lander, E. S. and Kruglyak, L. (1995) Genetic dissection of complex traits: Guidelines for interpreting and reporting linkage results. *Nat. Genet.* **11**: 241–247.

Legarra, A., Aguilar, I., and Misztal, I. (2009) A relationship matrix including full pedigree and genomic information. *J. Dairy Sci.* **92**: 4656–4663.

Legarra, A., Robert-Granie, C., Croiseau, P., Guillaume, F., and Fritz, S. (2011) Improved LASSO for genomic selection. *Genet. Res.* **93**: 77–87.

Lewontin, R. C. and Hubby, J. L. (1966) A molecular approach to the study of genic heterozygosity in natural populations. II. Amount of variation and degree of heterozygosity in natural populations of *Drosophila pseudoobscura*. *Genetics* **54**: 595–609.

Li, Y., Willer, C. J., Ding, J., Scheet, P., and Abecasis, G. R. (2010) MaCH: Using sequence and genotype data to estimate haplotypes and unobserved genotypes. *Genet. Epidemiol.* **34**: 816–834.

Lillehammer, M., Meuwissen, T. H. E., and Sonesson, A. K. (2011) A comparison of dairy cattle breeding designs that use genomic selection. *J. Dairy Sci.* **94**: 493–500.

Lin, C. Y. (1978) Index selection for genetic improvement of quantitative characters. *Theor. Appl. Genet.* **52**: 49–56.

Litt, M. and Luty, J. A. (1989) A hypervariable microsatellite revealed by in vitro amplification of a dinucleotide repeat within the cardiac muscle actin gene. *Am. J. Hum. Genet.* **44**: 397.

Lourenco, D. A. L., Misztal, I., Tsuruta, S., Aguilar, I., Ezra, E., Ron, M., Shirak, A., and Weller, J. I. (2014a) Methods for genomic evaluation in a relatively small dairy population and the impact of inclusion of genotyped cows' information in multiple-parity analyses. *J. Dairy Sci.* **97**: 1742–1752.

Lourenco, D. A. L., Misztal, I., Tsuruta, S., Aguilar, I., Lawlor, T. J., Forni, S., and Weller, J. I. (2014b) Are evaluations on young genotyped animals benefiting from the past generations? *J. Dairy Sci.* **97**: 3930–3942.

Lynch, M. and Walsh, B. (1998) *Genetics and Analysis of Quantitative Traits*. Sinauer Associates, Inc., Sunderland, MA.

Maccluer, J. W., Vandeberg, J. L., Read, B., and Ryder, O. A. (1986) Pedigree analysis by computer simulation. *Zoo Biol.* **5**: 147–160.

Mackay, T. F. (2001) The genetic architecture of quantitative traits. *Annu. Rev. Genet.* **35**: 303–339.

Mackinnon, M. J. and Georges, M. A. J. (1998) Marker-assisted preselection of young diary sires prior to progeny-testing. *Livest. Prod. Sci.* **54**: 229–250.

Mackinnon, M. J. and Weller, J. I. (1995). Methodology and accuracy of estimation of quantitative trait loci parameters in a half-sib design using maximum likelihood. *Genetics* **141**: 755–770.

Maher, B. (2008) Personal genomes: The case of the missing heritability. *Nature* **456**: 18–21.

Mangin, B., Goffinet, B., and Rebai, A. (1994) Constructing confidence intervals for QTL location. *Genetics* **138**: 1301–1308.

Marchini, J., Howie, B., Myers, S., McVean, G., and Donnelly, P. (2007) A new multipoint method for genome-wide association studies by imputation of genotypes. *Nat. Genet.* **39**: 906–913.

Martinez, O. and Curnow, R. N. (1992) Estimating the locations and the sizes of the effects of quantitative trait loci using flanking markers. *Theor. Appl. Genet.* **85**: 480–488.

Martinez, O. and Curnow, R. N. (1994) Missing markers when estimating quantitative trait loci using regression mapping. *Heredity* **73**: 198–206.

Matukumalli, L. K., Lawley, C. T., Schnabel, R. D., Taylor, J. F., Allan, M. F., Heaton, M. P., O'Connell, J., Moore, S. S., Smith, T. P. L., Sonstegard, T. S., and Van Tassell, C. P. (2009) Development and characterization of a high density SNP genotyping assay for cattle. *PLoS One* **4**: e5350.

McClure, M. C., Bickhart, D., Null, D., VanRaden, P., Xu, L., Wiggans, G., Liu, G., Schroeder, S., Glasscock, J., Armstrong, J., Cole, J. B., Van Tassell, C. P., and Sonstegard, T. S. (2014) Bovine exome sequence analysis and targeted SNP genotyping of recessive fertility defects BH1, HH2, and HH3 reveal a putative causative mutation in SMC2 for HH3. *PLoS One* **9**: e92769.

McVean, G. A., Abecasis, D. M., Auton, R. M., Brooks, G. A. R., Depristo, D. R., Durbin, A., Handsaker, A. G., Kang, P., Marth, E. E., McVean, P., Gabriel, S. B., Gibbs, R. A., Green, E. D., Hurles, M. E., Knoppers, B. M., Korbel, J. O., Lander, E. S., Lee, C., Lehrach, H., Mardis, E. R., Marth, G. T., McVean, G. A., Nickerson, D. A., Schmidt, J. P., Sherry, S. T., Wang, J., Wilson, R. K., Gibbs, R. A., Dinh, H., and Kovar, C. (2012) An integrated map of genetic variation from 1,092 human genomes. *Nature* **491**: 56–65.

Meuwissen, T. H. E. and Goddard, M. E. (2000) Fine mapping of quantitative trait loci using linkage disequilibria with closely linked marker loci. *Genetics* **155**: 421–430.

Meuwissen, T. H. E. and Goddard, M. E. (2004) Mapping multiple QTL using linkage disequilibrium and linkage analysis information and multitrait data. *Genet. Sel. Evol.* **36**: 261–279.

Meuwissen, T. H. E., Hayes, B. J., and Goddard, M. E. (2001) Prediction of total genetic value using genome-wide dense marker maps. *Genetics* **157**: 1819–1829.

Meuwissen, T. H. E., Karlsen, A., Lien, S., Olsaker, I., and Goddard, M. E. (2002) Fine mapping of a quantitative trait locus for twinning rate using combined linkage and linkage disequilibrium mapping. *Genetics* **161**: 373–379.

Misztal, I. and Wiggans, G. R. (1988) Approximation of prediction error variance in large-scale animal models. *J. Dairy Sci.* **71**(Suppl. 2): 27.

Misztal, I., Lawlor, T. J., Short, T. H., and Wiggans, G. R. (1991) Continuous genetic evaluation of Holsteins for type. *J. Dairy Sci.* **74**: 2001–2009.

Misztal, I., Legarra, A., and Aguilar, I. (2009) Computing procedures for genetic evaluation including phenotypic, full pedigree, and genomic information. *J. Dairy Sci.* **92**: 4648–4655.

Misztal, I., Aguilar, I., Legarra, A., and Lawlor, T. J. (2010) Choice of parameters for single-step genomic evaluation for type. *J. Dairy Sci.* **93**(Suppl. 1): 533.

Misztal, I., Tsuruta, S., Aguilar, I., Legarra, A., VanRaden, P. M., and Lawlor, T. J. (2013) Methods to approximate reliabilities in single-step genomic evaluation. *J. Dairy Sci.* **96**: 647–654.

Moav, R. (1966) Specialized sire and dam lines. I. Economic evaluation of crossbreeds. *Anim. Prod.* **8**: 193–202.

Moav, R. (1973) Economic evaluation of genetic difference. In: *Agricultural Genetics, Selected Topics*. R.Moav (Ed.) John Wiley & Sons, Inc., New York, NY, pp. 319–352.

Morgan, T. H. (1910) Chromosomes and heredity. *Am. Nat.* **44**: 194.

Moser, G., Khatkar, M. S., Hayes, B. J., and Raadsma, H. W.(2010) Accuracy of direct genomic values in Holstein bulls and cows using subsets of SNP markers. *Genet. Sel. Evol.* **42**: 37.

Mosig, M. O., Lipkin, E., Khutoreskaya, G., Tchourzyna, E., Soller, M., and Friedmann, A. (2001) A whole genome scan for quantitative trait loci affecting milk protein percentage in Israeli-Holstein cattle, by means of selective milk DNA pooling in a daughter design, using an adjusted false discovery rate criterion. *Genetics* **157**: 1683–1698.

Moskvina, V. and Schmidt, K. M. (2008) On multiple-testing correction in genome-wide association studies. *Genet. Epidemiol.* **32**: 567–573.

Mullis, K., Faloona, F., Scharf, S., Saiki, R., Horn, G., and Erlich, H. (1986) Specific enzymatic amplification of DNA in vitro: The polymerase chain reaction. *Cold Spring Harb. Symp. Quant. Biol.* **51**: 263–273.

Nadesalingam, J., Plante, Y., and Gibson, J. P. (2001) Detection of QTL for milk production on Chromosomes 1 and 6 of Holstein cattle. *Mamm. Genome* **12**: 27–31.

Nakaya, A. and Isobe, N. (2012) Will genomic selection be a practical method for plant breeding? *Ann. Bot.* **110**: 1301–1316.

Neimann-Sørensen, A. and Robertson, A. (1961) The association between blood groups and several production characters in three Danish cattle breeds. *Acta Agric. Scand.* **11**: 163–196.

Nemir, M., Bhattacharyya, D., Li, X., Singh, K., Mukherjee, A. B., and Mukherjee, B. B. (2000) Targeted inhibition of osteopontin expression in the mammary gland causes abnormal morphogenesis and lactation deficiency. *J. Biol. Chem.* **275**: 969–976.

Nothnagel, M., Ellinghaus, D., Schreiber, S., Krawczak, M., and Franke, A. (2009) A comprehensive evaluation of SNP genotype imputation. *Hum. Genet.* **125**: 163–171.

Nyholt, D. R. (2004) A simple correction for multiple testing for single-nucleotide polymorphisms in linkage disequilibrium with each other. *Am. J. Hum. Genet.* **74**: 765–769.

Ober, U., Erbe, M., Long, N., Porcu, E., Schlather, M., and Simianer, H. (2011) Predicting genetic values: A kernel-based best linear unbiased prediction with genomic data. *Genetics* **188**: 695–708.

Olsen, H. G., Gomez-Raya, L., Vage, D. I., Olsaker, I., Klungland, H., Svendsen, M., Adnoy, T., Sabry, A., Klemetsdal, G., Schulman, N., Kramer, W., Thaller, G., Ronningen, K., and Lien, S. (2002) A genome scan for quantitative trait loci affecting milk production in Norwegian dairy cattle. *J. Dairy Sci.* **85**: 3124–3130.

Olsen, H. G., Lien, S., Gautier, M., Nilsen, H., Roseth, A., Berg, P. R., Sundsaasen, M., Svendsen, K. K., and Meuwissen, T. H. (2005) Mapping of a milk production QTL to a 420 kb region on bovine chromosome 6. *Genetics* **169**: 275–283.

Olsen, H. G., Nilsen, H., Hayes, B., Berg, P. R., Svendsen, M., Lien, S., and Meuwissen, T. (2007) Genetic support for a quantitative trait nucleotide in the ABCG2 gene affecting milk composition of dairy cattle. *BMC Genet.* **8**: 32.

Owen, J. B. (1975) Selection of dairy bulls on half-sister records. *Anim. Prod.* **20**: 1–10.

Paterson, A. H., Lander, E. S., Hewitt, J. D., Peterson, S., Lincoln, S. E., and Tanksley, S. D. (1988) Resolution of quantitative traits into Mendelian factors by using a complete linkage map of restriction fragment length polymorphisms. *Nature* **335**: 721–726.

Patry, C. and Ducrocq, V. (2010) An approach to account for selection bias in national evaluations due to genomic selection. *Proceedings of the 9th World Congress Genetics Applied Livestock Production,* No. 0559, Leipzig, Germany.

Payne, F. (1918) The effect of artificial selection on bristle number in *Drosophila ampelophila* and its interpretation. *Proc. Natl. Acad. Sci. U. S. A.* **4**: 55–58.

Pei, Y.-F., Li, J., Zhang, L., Papasian, C. J., and Deng, H.-W. (2008) Analyses and comparison of accuracy of different genotype imputation methods. *PLoS One* **3**: e3551.

Pemberton, J. M. (2008) Wild pedigrees: The way forward. *Proc. Biol. Sci.* **275**: 613–621.

Pintus, M. A., Gaspa, G., Nicolazzi, E. L., Vicario, D., Rossoni, A., Ajmone-Marsan, P., Nardone, A., Dimauro, C., and Macciotta, N. P. P. (2012) Prediction of genomic breeding values for dairy traits in Italian Brown and Simmental bulls using a principal component approach. *J. Dairy Sci.* **95**: 3390–3400.

Přibyl, J., Madsen, P., Bauer, J., Přibylová, J., Šimečková, M., Vostrý, L., and Zavadilová, L. (2013) Contribution of domestic production records, Interbull estimated breeding values, and single nucleotide polymorphism genetic markers to the single-step genomic evaluation of milk production. *J. Dairy Sci.* **96**: 1865–1873.

Pryce, J. E. and Daetwyler, H. D. (2012) Designing dairy cattle breeding schemes under genomic selection: A review of international research. *Anim. Prod. Sci.* **52**: 107–114.

Purcell, S., Neale, B., Todd-Brown, K., Thomas, L., Ferreira, M. A. R., Bender, D., Maller, J., Sklar, P., de Bakker, P. I. W., Daly, M. J., and Sham, P. C.(2007) PLINK: A tool set for whole-genome association and population-based linkage analyses. *Am. J. Hum. Genet.* **81**: 559–575.

Qanbari, S., Pimentel, E. C. G., Tetens, J., Thaller, G., Lichtner, P., Sharifi, A. R., and Simianer, H. (2010) The pattern of linkage disequilibrium in German Holstein cattle. *Anim. Genet.* **41**: 346–356.

Qin, Z. H. S., Niu, T. H., and Liu, J. S. (2002) Partition-ligation-expectation-maximization algorithm for haplotype inference with single-nucleotide polymorphisms. *Am. J. Hum. Genet.* **71**: 1242–1247.

Quaas, R. L. and Pollak, E. J. (1980) Mixed model methodology for farm and ranch beef cattle testing programs. *J. Anim. Sci.* **51**: 1277–1287.

Ranade, K., Chang, M. S., Ting, C. T., Pei, D., Hsiao, C. F., Olivier, M., Pesich, R., Hebert, J., Chen, Y.-D. I., Dzau, V. J., Curb, D., Olshen, R., Risch, N., Cox, D. R., and Botstein, D. (2001) High-throughput genotyping with single nucleotide polymorphisms. *Genome Res.* **11**: 1262–1268.

Redon, R., Ishikawa, S., Fitch, K. R., Feuk, L., Perry, G. H., *et al.* (2006) Global variation in copy number in the human genome. *Nature* **444**: 444–454.

Riquet, J., Coppieters, W., Cambisano, N., Arranz, J. J., Berzi, P., Davis, S. K., Grisart, B., Farnir, F., Karim, L., Mni, M., Simon, P., Taylor, J. F., Vanmanshoven, P., Wagenaar, D., Womack, J. E., and Georges, M. (1999) Fine-mapping of quantitative trait loci by identity by descent in outbred populations: Application to milk production in dairy cattle. *Proc. Natl. Acad. Sci. U. S. A.* **96**: 9252–9257.

Robinson, G. K. (1986) Group effects and computing strategies for models for estimating breeding values. *J. Dairy Sci.* **69**: 3106–3111.

Ron, M. and Weller, J. I. (2007) From QTL to QTN identification in livestock—"winning by points rather than knock-out": A review. *Anim. Genet.* **38**: 429–439.

Ron, M., Yoffe, O., Ezra, E., Medrano, J. F., and Weller, J. I. (1994) Determination of milk protein effects on production traits of Israeli Holsteins. *J. Dairy Sci.* **77**: 1106–1113.

Ron, M., Kliger, D., Feldmesser, E., Seroussi, E., Ezra, E., and Weller, J. I. (2001) Multiple QTL analysis of bovine chromosome 6 in the Israeli Holstein population by a daughter design. *Genetics* **159**: 727–735.

Ron, M., Feldmesser, E., Golik, M., Tager-Cohen, I., Kliger, D., Reiss, V., Domochovsky, R., Alus, O., Seroussi, E., Ezra, E., and Weller, J. I. (2004) A complete genome scan of the Israeli Holstein population for quantitative trait loci by a daughter design. *J. Dairy Sci.* **87**: 476–490.

Ron, M., Cohen-Zinder, M., Peter, C., Weller, J. I., and Erhardt, G. (2006) ABCG2 polymorphism in *Bos indicus* and *Bos taurus* cattle breeds. *J. Dairy Sci.* **89**: 4921–4923.

Sargolzaei, M., Schenkel, F. S., Jansen, G. B., and Schaeffer, L. R. (2008) Extent of linkage disequilibrium in Holstein cattle in North America. *J. Dairy Sci.* **91**: 2106–2117.

SAS Institute Inc. (1999) *SAS/STAT User's Guide, Version 8*. SAS Institute Inc., Cary, NC.

Sax, K. (1923) The association of size differences with seed-coat pattern and pigmentation in *Phaseeolus vulgaris*. *Genetics* **8**: 552–560.

Schaeffer, L. R. (2006) Strategy for applying genome-wide selection in dairy cattle. *J. Anim. Breed. Genet.* **123**: 218–223.

Scheet, P. and Stephens, M. (2006) A fast and flexible statistical model for large-scale population genotype data: Applications to inferring missing genotypes and haplotypic phase. *Am. J. Hum. Genet.* **78**: 629–644.

Schnabel, R. D., Kim, J. J., Ashwell, M. S., Sonstegard, T. S., Van Tassell, C. P., Connor, E. E, and Taylor, J. F. (2005) Fine-mapping milk production quantitative trait loci on BTA6: Analysis of the bovine osteopontin gene. *Proc. Natl. Acad. Sci. U. S. A.* **102**: 6896–6901.

Schrooten, C., Bovenhuis, H., Coppieters, W., and van Arendonk, J. A. M. (2000) Whole genome scan to detect quantitative trait loci for conformation and functional traits in dairy cattle. *J. Dairy Sci.* **83**: 795–806.

Schulman, N. F., Viitala, S. M., de Koning, D. J., Virta, J., Maki-Tanila, A., and Vilkki, J. H. (2004) Quantitative trait loci for health traits in Finnish Ayrshire cattle. *J. Dairy Sci.* **87**: 443–449.

Searle, S. R. (1982) *Matrix Algebra Useful for Statistics*. John Wiley & Sons, Inc., New York, NY.

Searle, S. R., Casella, G., and McCulloch, C. E. (1992) *Variance Components*. John Wiley & Sons, Inc., New York, NY.

Seroussi, E., Glick, G., Shirak, A., Ezra, E., Zeron, Y., Ron, M., and Weller, J. I. (2013) Maternity validation using sire-only BovineSNP50 BeadChip data. *Anim. Genet.* **44**: 754–757.

Shriner, D., Adeyemo, A., Chen, G., and Rotimi, C. (2010) Practical considerations for imputation of untyped markers in admixed populations. *Genet. Epidemiol.* **34**: 258–265.

Šidák, Z. K. (1967) Rectangular confidence regions for the means of multivariate normal distributions. *J. Am. Stat. Assoc.* **62**: 626–633.

Simeone, R., Misztal, I., Aguilar, I., and Legarra, A. (2011) Evaluation of the utility of diagonal elements of the genomic relationship matrix as a diagnostic tool to detect mislabelled genotyped animals in a broiler chicken population. *J. Anim. Breed. Genet.* **128**: 386–393.

Simes, R. J. (1986) An improved Bonferroni procedure for multiple tests of significance. *Biometrika* **73**: 751–754.

Smaragdov, M. G. (2013) Genomic selection of milk cattle. The practical application over five years. *Russ. J. Genet.* **49**: 1089–1097.

Smith, C. and Simpson, S. P. (1986) The use of genetic polymorphisms in livestock improvement. *J. Anim. Breed. Genet.* **103**: 205–217.

Soller, M. (1994) Marker assisted selection—an overview. *Anim. Biotechnol.* **5**: 193–207.

Soller, M., Genizi, A., and Brody, T. (1976) On the power of experimental designs for the detection of linkage between marker loci and quantitative loci in crosses between inbred lines. *Theor. Appl. Genet.* **47**: 35–59.

Solomon, E. and Bodmer, W. F. (1979) Evolution of a sickle variant gene. *Lancet* **1**: 923.

Sørensen, A. C. and Sørensen, M. K. (2009) Genotyping both males and females is favourable in genomic dairy cattle breeding schemes. *Interbull Bull.* **40**: 94–97.

Southey, B. R. and Fernando, R. L. (1998) Controlling the proportion of false positives among significant results in QTL detection. *Proceedings of the 6th World Congress Genetics Applied Livestock Production*, Armidale, NSW, Australia, **26**: 221–224.

Spelman, R. J., Coppieters, W., Karim, L., van Arendonk, J. A. M., and Bovenhuis, H. (1996) Quantitative trait loci analysis for five milk production traits on chromosome six in the Dutch Holstein-Friesian population. *Genetics* **144**: 1799–1808.

Stormont, C. (1967) Contribution of blood typing to dairy science progress. *J. Dairy Sci.* **50**: 253–260.

Stormont, C., Owen, R. D., and Irwin, M. R. (1951) The B and C systems of bovine blood groups. *Genetics* **36**: 134–161.

Strickberger, M. W. (1969) *Genetics*. The Macmillan Company, New York, NY.

Sutton, W. S. (1903) The chromosomes in heredity. *Biol. Bull.* **4**: 213–251.

Tanksley, S. D., Medina-Filho, H., and Rick, C. M.(1982) Use of naturally occurring enzyme variation to detect and map genes controlling quantitative trait in an interspecific backcross of tomato. *Heredity* **49**: 11–25.

Tautz, D. (1989) Hypervariability of simple sequences as a general source of polymorphic DNA markers. *Nucleic Acids Res.* **17**: 6463–6471.

Thompson, R. (1979) Sire evaluations. *Biometrics* **35**: 339–353.

Thomsen, B., Horn, P., Panitz, F., Bendixen, E., Petersen, A. H., Holm, L.-E., Nielsen, V. H., Agerholm, J. S., Arnbjerg, J., and Bendixen, C. (2005) A missense mutation in the bovine SLC35A3 gene, encoding a UDP-N-acetylglucosamine transporter, causes complex vertebral malformation. *Genome Res.* **16**: 97–105.

Tibshirani, R. (1996) Regression shrinkage and selection via the Lasso. *J. R. Stat. Soc. Ser. B* **58**: 267–288.

Titterington, D. M., Smith, A. F. M., and Makov, U. E. (1985) *Statistical Analysis of Finite Mixture Distributions*. John Wiley & Sons, Inc., New York, NY.

Toro, M. A. and Varona, L. (2010) A note on mate allocation for dominance handling in genomic selection. *Genet. Sel. Evol.* **42**: 33.

Tsuruta, S., Misztal, I., and Strandén, I. (2001) Use of the preconditioned conjugate gradient algorithm as a generic solver for mixed-model equations in animal breeding applications. *J. Anim. Sci.* **79**: 1166–1172.

Tsuruta, S., Misztal, I., Aguilar, I., and Lawlor, T. J. (2011) Multiple-trait genomic evaluation of linear type traits using genomic and phenotypic data in US Holsteins. *J. Dairy Sci.* **94**: 4198–4204.

Ufford, G. R., Henderson, C. R, and VanVleck, L. D. (1979) Approximate procedure for determining prediction error variances of sire evaluations. *J. Dairy Sci.* **62**: 621–626.

Van der Vorst, H. (1992) Bi-CGSTAB: A fast and smoothly converging variant of Bi-CG for the solution of nonsymmetric linear systems. *SIAM J. Sci. Stat. Comput.* **13**: 631–644.

VanRaden, P. M. (2008) Efficient methods to compute genomic predictions. *J. Dairy Sci.* **91**: 4414–4423.

VanRaden, P. M. (2011) findhap.f90. Accessed July 9, 2012. http://aipl.arsusda.gov/software/findhap.

VanRaden, P. M. and Wiggans, G. R. (1991) Derivation, calculation and use of national animal model information. *J. Dairy Sci.* **74**: 2737–2746.

VanRaden, P. M., Van Tassell, C. P., Wiggans, G. R.,Sonstegard, T. S., Schnabel, R. D., Taylor, J. F., and Schenkel, F. S. (2009) Invited review: Reliability of genomic predictions for North American Holstein bulls. *J. Dairy Sci.* **92**: 16–24.

VanRaden, P. M., O'Connell, J. R., Wiggans, G. R., and Weigel, K. A. (2011a) Genomic evaluations with many more genotypes. *Genet. Sel. Evol.* **43**: 10.

VanRaden, P. M., Olson, K. M., Null, D. J., and Hutchison, J. L. (2011b) Harmful recessive effects on fertility detected by absence of homozygous haplotypes. *J. Dairy Sci.* **94**: 6153–6161.

VanRaden, P. M., Cooper, T. A., Wiggans, G. R., O'Connell, J. R., and Bacheller, L. R. (2013a) Confirmation and discovery of maternal grandsires and great-grandsires in dairy cattle. *J. Dairy Sci.* **96**: 1874–1879.

VanRaden, P. M., Null, D. J., Sargolzaei, M., Wiggans, G. R., Tooker, M. E., Cole, J. B., Sonstegard, T. S., Connor, E. E., Winters, M., van Kaam, J. B. C. H. M., Valentini, A., Van Doormaal, B. J., Faust, M. A., and Doak, G. A. (2013b) Genomic imputation and evaluation using high-density Holstein genotypes. *J. Dairy Sci.* **96**: 668–678.

Van Tassell, C. P., Casella, G., and Pollak, E. J. (1995) Effects of selection on estimates of variance components using Gibbs sampling and restricted maximum likelihood. *J. Dairy Sci.* **78**: 678–692.

Van Vleck, L. D. (1970a) Misidentification in estimating the paternal sib correlation. *J. Dairy Sci.* **53**: 1469–1474.

Van Vleck, L. D. (1970b) Misidentification and sire evaluation. *J. Dairy Sci.* **53**: 1697–1702.

Van Vleck, L. D. (1981) Potential genetic impact of artificial insemination, sex selection, embryo transfer, cloning, and selfing in dairy cattle. In: *New Technologies in Animal Breeding*. Academic Press, New York, NY, pp. 221–242.

Van Vleck, L. D. (1982) Is embryo transfer profitable? In: *Genetics Research 1981–1982 Report to Eastern Artificial Insemination Cooperative, Inc*. Ithaca, NY, pp. 88–100.

Vazquez, A. I., Rosa, G. J. M., Weigel, K. A., de los Campos, G., Gianola, D., and Allison, D. B. (2010) Predictive ability of subsets of single nucleotide polymorphisms with and without parent average in US Holsteins. *J. Dairy Sci.* **93**: 5942–5949.

Velmala, R. J., Vilkki, H. J., Elo, K. T., de Koning, D. J., and Maki-Tanila, A. V. (1999) A search for quantitative trait loci for milk production traits on chromosome 6 in Finnish Ayrshire cattle. *Anim. Genet.* **30**: 136–143.

Viitala, S. M., Schulman, N. F., de Koning, D. J., Elo, K., Kinos, R., Virta, A., Virta, J., Maki-Tanila, A., and Vilkki, J. H. (2003) Quantitative trait loci affecting milk production traits in Finnish Ayrshire dairy cattle. *J. Dairy Sci.* **86**: 1828–1836.

Vilkki, H. J., de Koning, D. J., Elo, K. T., Velmala, R., and Maki-Tanila, A. (1997) Multiple marker mapping of quantitative trait loci of Finnish dairy cattle by regression. *J. Dairy Sci.* **80**: 198–204.

Villanueva, B., Pong-Wong, R., Grundy, B., and Woolliams, J. A. (1999) Potential benefit from using an identified major gene in BLUP evaluation with truncation and optimal selection. *Genet. Sel. Evol.* **31**: 115–133.

Visscher, P. M. (2008) Sizing up human height variation. *Nat. Genet.* **40**: 489–490.

Visscher, P. M., Thompson, R., and Haley, C. S. (1996) Confidence intervals in QTL mapping by bootstrapping. *Genetics* **143**: 1013–1020.

Wang, H., Misztal, I., Aguilar, I., Legarra, A., and Muir, W. M. (2012) Genome-wide association mapping including phenotypes from relatives without genotypes. *Genet. Res.* **94**: 73–83.

Weber, J. L. and May, P. E. (1989) Abundant class of human DNA polymorphisms which can be typed using the polymerase chain reaction. *Am. J. Hum. Genet.* **44**: 388.

Weber, J. L. and Wong, C. (1993) Mutation of human short tandem repeats. *Hum. Mol. Genet.* **2**: 1123–1128.

Weigel, K. A., de los Campos, G., Gonzalez-Recio, O., Naya, H., Wu, X. L., Long, N., Rosa, G. J., and Gianola, D. (2009) Predictive ability of direct genomic values for lifetime net merit of Holstein sires using selected subsets of single nucleotide polymorphism markers. *J. Dairy Sci.* **92**: 5248–5257.

Weigel, K. A., Van Tassell, C. P., O'Connell, J. R., VanRaden, P. M., and Wiggans, G. R. (2010) Prediction of unobserved single nucleotide polymorphism genotypes of Jersey cattle using reference panels and population-based imputation algorithms. *J. Dairy Sci.* **93**: 2229–2238.

Weinberg, W. (1908). Über den Nachweis der Vererbung beim Menschen. *Jahreshefte des Vereins für vaterländische Naturkunde in Württemberg* **64**: 368–382. [About evidence of heredity in humans. Jahreshefte the association for patriotic Natural History in Württemberg.]

Weller, J. I. (1994) *Economic Aspects of Animal Breeding*. Chapman & Hall, London, 244 pp.

Weller, J. I. (2007) Marker assisted selection in dairy cattle. In: *Marker-Assisted Selection, Current Status and Future Perspectives in Crops, Livestock, Forestry and Fish*. E.Guimarães, J.Ruane, B. D.Scherf, A.Sonnino, and J. D.Dargie (Eds.) Food and Agriculture Organization of the United Nations, Rome, Italy, pp. 199–228.

Weller, J. I. (2009) *Quantitative Trait Loci Analysis in Animals*, 2nd edition. CABI Publishing, London, 272 pp.

Weller, J. I. and Ron, M. (2011) *Invited review:* Quantitative trait nucleotide determination in the era of genomic selection. *J. Dairy Sci.* **94**: 1082–1090.

Weller, J. I. and Soller, M. (1981) Methods for production of multimarker strains. *Theor. Appl. Genet.* **59**: 73–78.

Weller, J. I., Soller, M., and Brody, T. (1988) Linkage analysis of quantitative traits in an interspecific cross of tomato (*L. esculentum* × *L. pimpinellifolium*) by means of genetic markers. *Genetics* **118**: 329–339.

Weller, J. I., Kashi, Y., and Soller, M. (1990) Power of "daughter" and "granddaughter" designs for genetic mapping of quantitative traits in dairy cattle using genetic markers. *J. Dairy Sci.* **73**: 2525–2537.

Weller, J. I., Song, J. Z., Heyen, D. W., Lewin, H. A., and Ron, M. (1998) A new approach to the problem of multiple comparisons in the genetic dissection of complex traits. *Genetics* **150**: 1699–1706.

Weller, J. I., Feldmesser, E., Golik, M., Tager-Cohen, I., Domochovsky, R., Alus, O., Ezra, E., and Ron, M. (2004) Factors affecting incorrect paternity assignment in the Israeli Holstein population. *J. Dairy Sci.* **87**: 2627–2640.

Weller, J. I., Shlezinger, M., and Ron, M. (2005) Correcting for bias in estimation of quantitative trait loci effects. *Genet. Sel. Evol.* **37**: 501–522.

Weller, J. I., Glick, G., Ezra, E., Zeron, Y., Seroussi, E., and Ron, M. (2010) Paternity validation and estimation of genotyping error rate for the BovineSNP50 BeadChip. *Anim. Genet.* **41**: 551–553.

Weller, J. I., VanRaden, P. M., and Wiggans, G. R. (2013) Application of a posteriori granddaughter and modified granddaughter designs to determine Holstein haplotype effects. *J. Dairy Sci.* **96**: 5376–5387.

Weller, J. I., Cole, J. B., VanRaden, P. M., and Wiggans, G. R. (2014a) Application of the a posteriori granddaughter design to the Holstein genome. *Animal* **88**: 511–519.

Weller, J. I., Glick, G., Shirak, A., Ezra, E., Seroussi, E., Shemesh, M., Zeron, Y., and Ron, M. (2014b) Predictive ability of selected subsets of single nucleotide polymorphisms (SNPs) in a moderately sized dairy cattle population. *Animal* **8**: 208–216.

Wellmann, R. and Bennewitz, J. (2012) Bayesian models with dominance effects for genomic evaluation of quantitative traits. *Genet. Res.* **94**: 21–37.

Weng, Z., Zhang, Zh., Zhang, Q., Fu, W., He, S., and Ding, X. (2013) Comparison of different imputation methods from low- to high-density panels using Chinese Holstein cattle. *Animal* **7**: 729–735.

Westell, R. A., Quaas, R. L., and Van Vleck, L. D. (1988) Genetic groups in an animal model. *J. Dairy Sci.* **71**: 1310–1318.

Wiggans, G. R., Sonstegard, T. S., VanRaden, P. M., Matukumalli, L. K., Schnabel, R. D., Taylor, J. F., Schenkel, F. S., and Van Tassell, C. P. (2009) Selection of single-nucleotide polymorphisms and quality of genotypes used in genomic evaluation of dairy cattle in the United States and Canada. *J. Dairy Sci.* **92**: 3431–3436.

Wiggans, G. R., VanRaden, P. M., and Cooper, T. A. (2011) The genomic evaluation system in the United States: Past, present, future. *J. Dairy Sci.* **94**: 3202–3211.

Wiggans, G. R., Cooper, T. A., VanRaden, P. M., Olson, K. M., and Tooker, M. E. (2012a) Use of the Illumina Bovine3K BeadChip in dairy genomic evaluation. *J. Dairy Sci.* **95**: 1552–1558.

Wiggans, G. R., VanRaden, P. M., and Cooper, T. A. (2012b) *Technical note:* Adjustment of all cow evaluations for yield traits to be comparable with bull evaluations. *J. Dairy Sci.* **95**: 3444–3447.

Wiggans, G. R., Cooper, T. A., Van Tassell, C. P., Sonstegard, T. S., and Simpson, E. B. (2013) *Technical note:* Characteristics and use of the Illumina Bovine LD and GeneSeek Genomic Profiler low-density bead chips for genomic evaluation. *J. Dairy Sci.***96**: 1258–1263.

Winter, A., Kramer, W., Werner, F. A. O., Kollers, S., Kata, S., Durstewitz, G., Buitkamp, J., Womack, J. E., Thaller, G., and Fries, R. (2002) Association of a lysine-232/alanine polymorphism in a bovine gene encoding acyl-CoA: Diacylglycerol acyltransferase (DGAT1) with variation at a quantitative trait locus for milk fat content. *Proc. Natl. Acad. Sci. U. S. A.* **99**: 9300–9305.

Yang, J., Benyamin, B., McEvoy, B. P., Gordon, S., Henders, A. K., Nyholt, D. R., Madden, P. A., Heath, A. C., Martin, N. G., Montgomery, G. W., Goddard, M. E., and Visscher, P. M. (2010) Common SNPs explain a large proportion of heritability for human height. *Nat. Genet.* **42**: 565–569.

Yule, G. U. (1906) On the theory of inheritance of quantitative compound characters on the basis of Mendels laws—a preliminary note. *Report of the 3rd International Conference on Genetics*, Royal Horticultural Society, London. pp. 140–142.

Zeng, J., Toosi, A., Fernando, R. L., Dekkers, J. C. M., and Garrick, D. J. (2013) Genomic selection of purebred animals for crossbred performance in the presence of dominant gene action. *Genet. Sel. Evol.* **45**: 11.

Zhang, W. and Smith, C. (1992) Computer simulation of marker-assisted selection utilizing linkage disequilibrium. *Theor. Appl. Genet.* **83**: 813–820.

Zhang, W. and Smith, C. (1993) Simulation of marker-assisted selection utilizing linkage disequilibrium: The effects of several additional factors. *Theor. Appl. Genet.* **86**: 492–496.

Zhang, Q., Boichard, D., Hoeschele, I., Ernst, C., Eggen, A., Murkve, B., Pfister-Genskow, M., Witte, L. A., Grignola, F. E., Uimari, P., Thaller, G., and BishopM. D. (1998) Mapping quantitative trait loci for milk production and health of dairy cattle in a large outbred pedigree. *Genetics* **149**: 1959–1973.

Zhang, Z., Ding, X., Liu, J., Zhang, Q., and de Koning, D. J. (2011) Accuracy of genomic prediction using low-density marker panels. *J. Dairy Sci.* **94**: 3642–3650.

Zou, H. (2006) The adaptive lasso and its oracle properties. *J. Am. Stat. Assoc.* **101**: 1418–1429.

Zou, H. and Hastie, T. (2005) Regularization and variable selection via elastic net. *J. R. Stat. Soc. Ser. B.* **67**: 301–320.

Index

Page numbers in *italics* refer to illustrations; those in **bold** refer to tables

Genomic Selection in Animals, First Edition. Joel Ira Weller.